Ground Combat Operations

I0160490

U.S. Marine Corps

PCN 139 000423 00

E R R A T U M

to

MCWP 3-1

GROUND COMBAT OPERATIONS

1. For administrative purposes, FMFM 6 is reidentified as MCWP 3-1.

DEPARTMENT OF THE NAVY
Headquarters United States Marine Corps
Washington, DC 20380-1775

4 April 1995

FOREWORD

1. PURPOSE

Fleet Marine Force Manual (FMFM) 6, *Ground Combat Operations,* provides the doctrinal basis for the planning and execution of ground combat operations for ground forces within the Marine air-ground task force (MAGTF). It establishes a common reference for operational and tactical operations routinely conducted by the ground combat element of the MAGTF.

FMFM 6 is the keystone manual for all subsequent ground combat-oriented manuals and is not intended to provide detailed tactics, techniques, and procedures for specific ground combat organizations. It provides the link between the tactics, techniques, and procedures provided in 6-series manuals and FMFM 1, *Warfighting*; FMFM 1-1, *Campaigning*; FMFM 1-2, *The Role of the Marine Corps in the National Defense;* and FMFM 1-3, *Tactics*.

2. SCOPE

FMFM 6 is designed for all Marine leaders regardless of military occupational specialty. This manual concentrates primarily on the ground combat element's warfighting capabilities, as the decisive maneuver force of the MAGTF, in the conduct of ground combat including maritime expeditionary operations and subsequent and sustained operations ashore. Its content pertains equally to combat support and combat service support organizations.

3. SUPERSESSION

OH 6-1, *Ground Combat Operations*, dated January 1988.

4. CHANGES

Recommendations for improving this manual are invited from commands as well as directly from individuals. Forward suggestions using the User Suggestion Form format to —

Commanding General
Doctrine Division (C 42)
Marine Corps Combat Development Command
2042 Broadway Street, Suite 210
Quantico, Virginia 22134-5021

5. CERTIFICATION

Reviewed and approved this date.

BY DIRECTION OF THE COMMANDANT OF THE MARINE CORPS

C. E. WILHELM
Lieutenant General, U. S. Marine Corps
Commanding General
Marine Corps Combat Development Command
Quantico, Virginia

DISTRIBUTION: 139 000423 00

Ground Combat Operations

Table of Contents

Chapter 4. Operational Maneuver From the Sea

Chapter 5. Offensive Operations

Chapter 6. Defensive Operations

Chapter 7. Operations Other Than War

Chapter 8. MAGTF GCE Operations in a Joint and Combined Environment

Glossary

Index

User Suggestion Form

From:

To: Commanding General
     Doctrine Division (C 42)
     Marine Corps Combat Development Command
     2042 Broadway Street Suite 210
     Quantico, Virginia 22134-5021

Subj: RECOMMENDATIONS CONCERNING FMFM 6, *GROUND*
     COMBAT OPERATIONS

1. In accordance with the foreword to FMFM 6, which invites
individuals to submit suggestions concerning this FMFM directly to
the above addressee, the following unclassified recommendation is
forwarded:

Page No.	Article/Paragraph No.	Line No.	Figure/Table

Nature of Change: ☐ Add
                   ☐ Delete
                   ☐ Change
                   ☐ Correct

2.   Proposed new verbatim text: (Verbatim, double-spaced; continue on additional pages as necessary.)

3.   Justification/source: (Need not be double-spaced.)

Note: Only one recommendation per page.

Record of Changes

(reverse blank)

Chapter 1

Ground Combat Doctrine

Marine ground forces must be prepared to conduct operations across the spectrum of war and in any operational environment. These operations can range from those conducted in war to actions that support peace. As the number and nature of operations change and expand, the role of the ground combat element (GCE) also expands and evolves. No longer can the GCE expect to be the supported element or designated the main effort during all phases of an operation. The expanded role of the GCE may require the GCE commander to accomplish his assigned mission with fewer resources than traditionally provided by the MAGTF. As a result, the demands placed on the GCE will be greater than ever before.

FMFM 6, *Ground Combat Operations,* provides the doctrinal foundation for the expanded role of the GCE when employed in MAGTF operations. It is critical to understanding Marine Corps doctrine for ground combat and the role of the GCE as part of the combined arms MAGTF. This manual provides connectivity between the philosophy of employing the MAGTF and the tactics, techniques, and procedures disseminated by the subordinate 6-series manuals.

The Marine Air-Ground Task Force

Understanding of the principles of war and the fundamentals of ground combat has led to the Marine Corps' adoption of maneuver warfare. Maneuver warfare is a warfighting philosophy that seeks to shatter the enemy's cohesion through a series of rapid, violent, and unexpected actions which create a turbulent and rapidly deteriorating situation with which the enemy cannot cope. The MAGTF is the force that the Marine Corps employs to conduct maneuver warfare in a combined arms operation. The nature of the MAGTF—cohesion, unity of effort, flexibility, and self-sustainment—makes it equal to the requirements of combined arms warfare.

The MAGTF contains four elements that can be tailored to a combined arms operation: a *command element*, a *ground combat element*, an *aviation combat element,* and *a combat service support element*. The MAGTF draws forces from ground, aviation, and combat service support organizations of the Fleet Marine Force (FMF) to meet this requirement.

Command Element

The command element (CE) is comprised of the MAGTF commander and his principal staff, detachments from the Marine expeditionary force (MEF) and the Marine Force (MARFORPAC and MARFORLANT). The CE provides the catalyst for unity of effort. It contains a command and control (C^2) system for the effective planning and execution of operations and is capable of exercising the requirements of joint force command and control warfare (C^2W).

Ground Combat Element

The ground combat element (GCE) is organized from resources and units of one or more Marine divisions. This includes the division headquarters, the infantry regiments, the artillery regiment(s), and separate battalions. Resources from each are weighed based on the capability that it can provide to the force and the task assigned or anticipated. In some cases, it may be necessary to draw from the selected Marine Corps reserve, joint, or combined ground combat units to meet the requirements of the mission.

Aviation Combat Element

The aviation combat element (ACE) is task organized from the Marine aircraft wing. *The ACE is a combat arm of the MAGTF, not a supporting arm of the GCE.* The MAGTF commander uses aviation's inherent reach not only for the direct application of firepower, but also to extend the range of the GCE. The MAGTF commander employs the ACE to—

- Fix the enemy to allow another force to maneuver to advantage and destroy him.
- Destroy an enemy force fixed by another force.
- Fix and destroy an enemy force.
- Attack specific enemy capabilities to shape the future decisive operation.

Combat Service Support Element

The combat service support element (CSSE) is task organized from the force service support group. It is tailored to meet specific support requirements of the MAGTF to provide timely, reliable, and continuous combat service support (CSS). The functional areas of CSS are supply, maintenance, transportation, general engineering, health service, and services (disbursing, postal, exchange, legal, etc). The CSS for MAGTF aviation units is provided by tailored Marine wing support group assets located in the ACE.

The Employment of the GCE

Marine ground combat is *always* viewed through the employment of the MAGTF. *The GCE is an integral component of the MAGTF and MAGTF operations. It is not employed as an independent entity.* The GCE, as does the ACE and CSSE, receives its mission from the MAGTF commander and accomplishes this mission in support of the MAGTF commander's intent. The coordination of MAGTF subordinate element operations on the uncertain battlefield is accomplished through unanimous understanding of the MAGTF commander's intent, his designation of the main effort, and the orchestration of MAGTF assets in time and space to accomplish battlefield tasks.

The GCE is capable of maneuvering to advantage against the enemy and applying in combination direct and indirect fires against him. However, the synergism of the MAGTF greatly enhances the combined arms capabilities resident in the GCE by extending the battlespace through—

- Application of firepower.
- Intelligence gathering and assimilation.
- Target acquisition.
- Mobility.

Furthermore, to the GCE commander, the appreciation of the MAGTF as a whole cannot be lost as, by its very nature, the GCE is limited in a number of functional areas. *The GCE and the other major subordinate elements of the MAGTF are intradependent. That is to say that the MAGTF is greater than the sum of its parts.* Each relies heavily on the capabilities of the others to maximize its own capabilities and lethality. The GCE must rely heavily on the ACE for deep and close supporting fires, mobility, and extended battlefield vision. The CSSE provides sustainment, beyond the organic CSS within the GCE, to ensure continuous operations and freedom of action of the GCE. The CE focuses the operations of all elements toward the successful accomplishment of the mission, task organizes the force, and enhances information connectivity between the elements.

A Doctrinal Foundation Based on the Principles of War

Changes in organization, armament, and employment of Marine Corps ground forces have been evolutionary. Likewise, the MAGTF, chosen as our organization for combat, has gradually changed to meet the needs of the modern battlefield. On the other hand, doctrinal changes have been revolutionary in nature in response to new threats, resources, or advances in technology. The distribution of *The Tentative Landing Manual* in 1933 and the issuance of FMFM 1, *Warfighting*, in 1989 illustrate this span of doctrine development. However, most ground combat operations can be assessed using historically developed truths governing the prosecution of war. These truths are the principles of war.

The Marine Corps' warfighting philosophy and organization for combat are based on the principles of war. The nine principles themselves are basic, but their application varies with the situation. The influencing factors are mission, enemy, terrain and weather, troops and support available-time available (METT-T). The application of the foregoing principles to the preparation for and overall direction of war is known as *strategy*. Their application to the conduct of a campaign is known as *operational art*. Their application to specific operations is called *tactics*. While the principles of war are unchanging, different combat situations exert varying effects which must be carefully considered if the principles are to be successfully applied. The proper application of these principles constitutes the true measure of military art.

Objective

Every successful military operation is directed toward a clearly defined, decisive, and attainable objective. The ultimate military objective of war is to defeat the enemy's forces or destroy his will to fight. At the strategic level of war, the objective is destruction of the enemy's center of gravity. *Center of gravity* is those characteristics, capabilities, or localities from which a military force derives its freedom of action, physical strength, or will to fight. At the operational and tactical levels of war, commanders aim to identify and attack critical vulnerabilities, which, when the attack is successful, can lead to the destruction of the enemy center of gravity. Subordinate unit objectives must contribute quickly and economically to the ultimate purpose of the operation.

The selection of an objective is based on consideration of METT-T. Every commander must clearly understand the overall mission and intent of the higher commander, his own mission, and the tasks he must perform. He considers every contemplated action in light of its direct contribution to the objective. He must communicate clearly to his subordinates the overall objective of the operation. The ability to select objectives whose attainment contributes most decisively and quickly to the defeat of the hostile armed forces is an essential attribute of an able commander.

The GCE commander directs his operations at enemy vulnerabilities. A vulnerability is a capability that is susceptible to attack. A critical vulnerability is a capability that is both susceptible to attack and critical to the enemy force's success. On the dynamic battlefield, identification of vulnerabilities is difficult, and, when they are discovered, the commander must quickly decide on the course of action he will pursue. Likewise, enemy vulnerabilities may become or cease to be critical for brief periods of time. The commander who can identify and take quick action against critical vulnerabilities dictates the tempo of operations.

Offensive

The offense alone brings victory; the defense can only avoid defeat. In taking the offensive, an attacker seizes, retains, and exploits the initiative and maintains freedom of action. The offense allows the commander to impose his will on the enemy, to determine the course of the battle, and to exploit enemy weaknesses. A defensive posture should be only a temporary expedient until the means are available to resume the offensive. Even in the conduct of a defense, the commander seeks every opportunity to seize the initiative by offensive action.

Offensive action can help the commander dictate the tempo of an operation. Our goal is to deprive the enemy of opportunities relevant to his operational objectives by putting him on a reactive footing. The GCE commander can accomplish this through swift decisionmaking coupled with rapid execution on the battlefield. The faster we can operate, the less time the enemy has to react to our actions and to plan actions of his own or according to General Patton, *". . .when we are attacking, the enemy has to parry, while, when we are defending or preparing to attack, he can attack us."*

Offensive action is not discouraged based on enemy advantages in troops and resources. Defensive operations against a vastly superior foe may only delay inevitable defeat. General Lee's attack at Chancellorsville and Rommel's operations in North Africa against numerically greater and better equipped armies illustrate the value of the offense at *every opportunity*.

> "I was too weak to defend, so I attacked."
> —General Robert E.
> Lee,

Mass

Combat power must be concentrated at the decisive place and time to achieve decisive results. Mass in ancient times meant sheer weight of numbers at a critical point. Today, mass means potential strength at the critical point or the ability to have it there before the enemy. Proper application of the principle of mass may achieve decisive local superiority for a numerically inferior force. Contributors to achieving mass include—

- Leadership.
- Troop strength.
- Tactical dispositions.
- Skillful use of fires.
- Combat support and CSS.
- Discipline, morale, and resolution.
- C^2.

> "The principles of war could, for brevity, be condensed into a single word— concentration."
> —B. H. Liddell Hart

The commander's attempt to mass is embodied in his *main effort*. The main effort is designed to successfully attack an enemy vulnerability or critical vulnerability. The main effort is a subordinate unit specifically designated by the commander that is given the preponderance of combat power and support to ensure success. All units and organizations must support the main effort. When the GCE is designated the main effort of the MAGTF, it must receive the support necessary for success. Though the MAGTF commander makes the ultimate decision regarding task organization of the force, the GCE commander must have the fortitude to ask for additional forces when the resources provided are inadequate to conduct the tasks assigned. This is important to the understanding of main effort, as the main effort must accomplish the mission assigned regardless of supporting effort failures.

The decision to concentrate a main effort requires strict economy and the acceptance of risk elsewhere. Due to the lethality of modern weapons, forces must be massed quickly and unexpectedly from dispersed formations and dispersed again after accomplishment of the mission. The commander concentrates forces and masses fires to exploit enemy weakness or where terrain offers the best opportunity to make maximum use of fire and maneuver. At the decisive place and time, the commander commits his reserve to generate the greatest combat power.

Economy of Force

Economy of force is the reciprocal of the principle of mass. This principle may be defined as the conservation of men and material in order that the maximum of fighting means will be available at the decisive time and place. The commander allocates the minimum essential combat power to exert pressure in secondary efforts and concentrates his greatest strength at the decisive point. This requires the acceptance of prudent risks in secondary areas to achieve superiority at the decisive place. Supporting efforts must directly support the main effort. Thus, forces not made available to the main effort are justifiable only when they divert superior enemy combat power from the decisive action or when they debilitate the enemy commander's decisionmaking ability.

Maneuver

Maneuver is the movement of forces on the battlefield in combination with fire or fire potential to achieve a position of advantage in respect to the enemy. Maneuver is an essential element of combat power. Maneuver in itself cannot produce decisive results. Combined with mass, offensive, economy of force, and surprise, maneuver provides favorable conditions for closing with the enemy. Maneuver contributes significantly to sustaining the initiative, exploiting success, preserving freedom of action, and reducing vulnerability. It is through maneuver that an inferior force can achieve decisive superiority at the necessary time and place. In many cases, maneuver is made possible only through the control of tempo and effective employment of firepower. The commander integrates supporting fires with the scheme of maneuver to create a dilemma for the enemy. Likewise, movement without fires exposes the force to effective enemy counteraction and risks losing the initiative and momentum. Maneuver that does not include violent action against the enemy will not be decisive. At all levels, successful application of this principle requires flexibility of thought and plans.

Unity of Command

Unity of command is the vesting of a single commander with the requisite authority to direct and coordinate the actions of all forces employed toward a common objective. Unity of command obtains the unity of effort that is essential to the decisive application of all available combat power. Subordinates are then focused on attaining the overall objectives as communicated from a **single commander**. In turn this fosters freedom of action, decentralized control, and initiative.

> "Nothing in war is so important as an undivided command."
>
> —Napoleon

Clearly stated intent and trust in subordinates by the commander are key to initiative and decentralized control. The commander's intent provides the force with an understanding of what the commander wants to do to the enemy and the desired end state. It is absolutely essential to unity of effort. Trust in subordinates is embodied in mission tactics. Mission tactics are initiated with mission-type orders. Mission orders are the assignment of missions with a clear task and purpose to a subordinate without dictating how to accomplish it.

Essential to maintaining unity is identification of the focus of effort. Of all the activities going on within the command, the commander recognizes the focus of effort as the most critical to success. The focus of effort is directed at that object or function which will cause the most decisive damage to the enemy. Normally, the main effort is assigned responsibility for accomplishing the focus of effort. It then becomes clear to all other units in the command that they must support the main effort. Like the commander's intent, the focus of effort becomes a harmonizing force.

Security

Security is achieved by those measures taken to prevent surprise, to ensure freedom of action, and to deny the enemy information about friendly forces. Security is essential to the protection of combat power; however, it does not imply overcautiousness or the avoidance of calculated risk. Adequate security against surprise requires a correct estimate of enemy capabilities, sufficient security measures, effective reconnaissance, and readiness for action. Security often is enhanced by bold seizure and retention of the initiative and speed, which denies the enemy the chance to interfere. Every unit

1-11

is responsible for its own local security, regardless of security measures implemented by a higher echelon.

Surprise

Surprise is the ability to strike the enemy at a time or place or in a manner for which he is unprepared. Surprise is a combat multiplier. He who can achieve it and can protect himself from it gains leverage. It is not essential that the enemy be taken unaware, but only that he become aware too late to react effectively. The effect of surprise is only temporary. To reap the benefits of surprise, the commander must exploit its initial shock, allowing the enemy no time to recover. An enemy taken unaware loses confidence in himself and his leaders, his morale drops, and he is, then, less able to take effective countermeasures. Surprise delays enemy reactions, overloads and confuses his C^2 systems, and provides initiative and momentum to the force.

By reducing enemy combat power, surprise enables a force to succeed with fewer forces than might otherwise be needed. Achieving outright surprise is never easy, especially with modern surveillance and warning capabilities. While always seeking surprise and being prepared to exploit it aggressively, the commander must also have a plan if surprise is lost. However, surprise can still be achieved by operating contrary to the enemy's expectations. Factors contributing to surprise include—

- Speed.
- Use of unexpected forces.
- Operating at night/during limited visibility.
- Effective and timely intelligence.
- Deception.
- Security.
- Variation in tactics and techniques.
- Use of terrain that appears unfavorable.

Simplicity

Plans should be as simple and direct as the accomplishment of the mission will permit. Direct, simple plans and clear, concise orders reduce the chance for misunderstanding and confusion, and they promote effective execution. Other factors being equal, the simplest plan is preferred.

Fundamentals of Ground Combat

The fundamentals of ground combat are general rules evolved from logical and time-proven application of the principles of war to both offensive and defensive combat.

Maintain Situational Awareness

The commander must be knowledgeable of the situation of his own force, that of the enemy force, the traits of the enemy commander, and the nature of the area of operations. The commander accomplishes this by locating and gaining contact with the enemy and constantly developing the situation. By gaining and maintaining contact, the commander is provided information about the enemy and is less vulnerable to surprise. Contact may vary from observation to close combat. Knowledge of the enemy's location, disposition, and movement is a potentially decisive advantage that must not be surrendered. Developing the situation consists of those actions taken to determine the strength, composition, and disposition of the enemy. This provides the commander with accurate information for developing his plan.

Exploit Known Enemy Gaps

The commander avoids enemy surfaces and attacks with maximum speed and combat power against enemy gaps. A gap is any weakness in the enemy force, not just physical preparations and employment. Gaps include—

- Poor morale.
- Tactical error.
- Lack of preparation.
- Lack of mutual support.
- Predictable operating patterns.

Control Key Terrain

The commander seeks to dominate key terrain that provides advantage of observation, cover and concealment, and fields of fire; that controls avenues of approach; and that provides security. In any zone of action or defensive sector, there are likely to be several key terrain features. The commander identifies them and plans to use them or to deny their use to the enemy. The possession of terrain is important only so far as the advantages it provides are exploited. Occupying terrain is not a goal in itself. The true purpose of an operation is defeat of the enemy. Terrain outside the zone of action that can dominate actions within the zone is also key terrain. The commander ensures control of this terrain through coordination with friendly adjacent units or by requesting that boundaries be moved to include this terrain within the zone or sector.

Dictate the Tempo of Operations

A paramount goal for the commander is to seize and retain the initiative in order to dictate the terms of the battle instead of having to react to the actions of the enemy. Aggressive employment of combat power, surprise, and exploitation of enemy errors all serve to gain or retain the initiative. The initiative normally belongs to the attacker at the beginning of an attack.

Neutralize the Enemy's Ability to React

The commander makes every effort to disrupt and degrade the enemy's ability to react to his plan. In so doing, the initiative is retained. Enemy capabilities are neutralized in depth by—

- Using fire support to shape the battlefield.
- Suppressing enemy forces and disrupting enemy support activities.
- Attacking the enemy's C^2.
- Isolating the battlefield and blocking enemy reinforcements.
- Concealing intentions from the enemy through proper security measures.
- Deceiving the enemy through the use of diversions and other techniques.

Maintain Momentum

Momentum is the increase of combat power, gained from seizing the initiative and attacking aggressively and rapidly. It is a function of initiative, concentration, and speed. Once an attack is launched, the commander makes every effort to build momentum until the attack becomes overwhelming and the mission is accomplished.

The commander does not sacrifice momentum to preserve the alignment of advancing units. He drives hard at those points offering the least resistance. The attacker does not waste combat power and time on enemy units that cannot jeopardize the overall mission, choosing instead to contain them with minimal forces and bypass them.

The defender gains momentum by employing the effects of his weapon systems in mass against the attacker's critical vulnerablities and by exploiting success gained through counterattack and rapid transition to the offense. Pressure against a weakening enemy must be relentless.

Act Quickly

Speed is essential to success. It promotes surprise, keeps the enemy off balance, contributes to the security of the force, makes the force a more difficult target, and prevents the enemy from taking effective countermeasures. Speed applies not only to physical movement but also to operational tempo, the exercise of command, staff functions, coordination, and all support activities. The commander who makes his estimate, decides on a course of action, and develops and executes his plan more quickly than his enemy counterpart can retain the initiative and dictate the conditions of the battle.

Speed can confuse and immobilize the enemy. It can compensate for a lack of mass and provide the momentum that the force requires. Attacking forces must move quickly to follow reconnaissance elements or successful probes through gaps in enemy defenses. The defender must recognize opportunities created by successful defense or enemy movement and rapidly counterattack. The commander must shift his main effort quickly to adjust to exploitable advantages; subordinate units must expect a shift in the main effort; and they must be be prepared to react appropriately.

The enemy must never be given the time to recover from the shock of the initial assault or counterattack, to identify the main effort, or to mass against the force.

Exploit Success

A successful attack or defense must be pressed relentlessly to prevent the enemy from recovering from the initial shock. Plans must provide for the exploitation of any advantage gained. When the **opportunity for decisive action** arises, the commander **commits his total resources** and demands the maximum effort from his troops. One of the most effective ways the commander can exploit success is by committing his reserve.

Be Flexible

The plan must foresee developments as far ahead as possible. However, it must also anticipate uncertainties and must be ready to exploit opportunities. The commander must be prepared to modify his plan and to shift his main effort in response to any situation. The commander maintains flexibility by retaining a balanced reserve, developing a simple plan, minimizing restrictions on subordinates, and immediately reconstituting a committed reserve.

Be Audacious

Audacity is the bold, intrepid, and aggressive execution of the operation, characterized by seizing every opportunity to strike a decisive blow against the enemy and relentlessly exploiting every success. Like speed and surprise, aggressive action enhances combat power beyond the material assets at hand.

Violence of action ensures success. All efforts to dictate tempo, maneuver to gain positional advantage, surprise the enemy, and apply combined arms against him will be jeopardized if the final effort against him is timidly executed. At all levels of the GCE, regardless of location on the battlefield, violence of action in the face of the enemy is required. Violence of action is an integral component of maneuver warfare. Violence against the enemy during the conduct of maneuver warfare is no different than that experienced in past wars and is not to be solely associated with an "attrition" style of warfare.

Provide for the Security of the Force

Security is always necessary, whether a force is assembling, on the march, or in combat. Security measures are dictated by the likelihood of contact with the enemy. Security is achieved by detecting the enemy and by providing time and space to react to the enemy. Rapid and aggressive offensive action provides a measure of inherent security.

The GCE conducts security operations in support of its own operations and may conduct these operations in support of other elements of the MAGTF or the MAGTF as a whole. When not designated the main effort, the GCE can expect to be tasked to provide security assistance to another element. However, security support must be provided voluntarily, without MAGTF tasking, when the situation demands GCE action. The GCE commander must not wait to be directed to provide this assistance. The GCE cannot wish away the security weaknesses of the other elements, and the other elements must do the utmost possible to limit GCE assistance requirements.

Task Organization of the GCE

The GCE is always task organized. Task organization is a temporary grouping of forces designed to accomplish a particular mission. Task organization is always based on METT-T. *Organization of the force without a METT-T basis is the antithesis of task organization.*

Commanders who ensure subordinate units receive an even distribution of support or equal share of assets, *despite the assignment of divergent tasks*, have given little thought to the prudent distribution of forces and the weighting of the main effort. For example, it is not uncommon to see each infantry unit receive tank support or a "fair share" of available combat engineers, antiarmor and heavy machinegun assets.

Habitual relationship is the customary combination of specific combat, combat support, and combat service support units, not task organization. Habitual relationships are valuable because a working relationship exists between the personalities of the organizations

involved, making coordination and anticipation of actions easier. Habitual relationships are also valuable since tactical procedures are well known by the affected units resulting in a greater capability to dictate the tempo of operations. Habitual relationships facilitate future task organization and supporting relationships such as that between artillery and infantry units. However, *until the mission is analyzed and a conscious decision* is made to attach certain units, a habitual relationship does not equate to task organization.

Chapter 2

GCE Combined-Arms Operations: What the GCE Brings to the Fight

Combined arms is the tactics, techniques, and procedures employed by a force to integrate firepower and mobility to produce a desired effect on the enemy. The GCE commander has the means to conduct combined arms operations. However, it is imperative that the GCE's resources be integrated with the full complement of MAGTF capabilities and brought to bear against the enemy. For the MAGTF commander, the GCE provides a capability to produce intelligence, conduct maneuver, apply firepower, and provide security.

Intelligence

Intelligence is fundamental to combat operations. Collection of information is the responsibility of every organization, unit, and individual and is a continuous activity. Lack of detailed information regarding enemy dispositions, capabilities, strengths, and weaknesses creates conditions that prevent effective combined arms. Incomplete intelligence weakens the coordination of firepower and maneuver against critical objectives, and situational awareness becomes an impossibility. However, it is in this very environment that the GCE must excel. The commander must expect that intelligence will never be complete even though sources of intelligence continue to expand. The commander cannot wait for complete intelligence to make a decision even though some critical information requirements are not fully met. The commander must prepare a flexible scheme of maneuver that will take advantage of the results of the intelligence collection efforts of his organic assets, as well as those of higher headquarters. Commanders not allocated specific intelligence collection assets must dedicate organic units to this function.

When task organized properly, the GCE has the capability to collect a variety of information within its area of influence. However, the GCE must integrate this information with other theater and national reporting to develop the enemy situation fully. This is particularly critical when planning for future operations. Close coordination must be conducted with the MAGTF CE and ACE to ensure assets organic to the CE, ACE, and those of higher/adjacent units are appropriately integrated to satisfy the commander's intelligence requirements.

Intelligence preparation of the battlespace (IPB), a component of the commander's preparation of the battlespace, is critical to the determination of the GCE's collection effort. It is a systematic and continuous process that analyzes the enemy, weather, and terrain in an area of operations. The IPB process integrates enemy doctrine and his mission with the effects of weather and terrain to evaluate his capabilities, vulnerabilities, and probable courses of action. The object or goal of the IPB is to support the commander's critical information requirements. The GCE commander bases his reconnaissance and surveillance (R&S) plan on the IPB. The R&S plan focuses on those critical areas for which information-gathering assets are the only means capable of providing that information due to the limited number and type of collection assets.

All units within the GCE have an inherent collection responsibility. To avoid under- or overreporting, the commander must articulate and disseminate those areas or items of information he deems necessary to his decisionmaking process. The GCE or MAGTF commander then tasks units that are specifically designed to gather information or units that have unique expertise (engineers, civil affairs, etc.) to collect information. Examples of these units are as follows:

- **Reconnaissance units** within the Marine division have the primary function of collection within the area of influence. They may be tasked to perform area or zone reconnaissance, amphibious reconnaissance, and surveillance. These units are most valuable when oriented on the enemy force as opposed to a GCE security orientation.

- The **scout-sniper platoon**, located in the infantry battalion, provides ground surveillance and scout snipers for specific assignments determined by the infantry battalion commander or in support of the higher GCE commander's and MAGTF commander's information requirements.

- **Light-armored reconnaissance units** provide the capability to conduct extended-range ground reconnaissance and surveillance for the GCE and MAGTF commander.

- **Engineers units** may conduct or assist in the conduct of area and zone reconnaissance operations. Engineer support to R&S operations may include road, bridge, and river crossing surveys and obstacle evaluation.

- **Forward observers, terminal controllers, and systems**, such as forward air controllers and target acquistion radars within artillery units, also provide sources of information, particularly for current operations when in direct contact with the enemy.

- **Infantry units** provide a continuous source of information through standard reporting, the maintenance of observation posts, sentinel posts, listening posts, and patrols. As reporting of enemy and terrain within their areas of responsibility may be critical to the larger scheme of maneuver, the significance of "routine" reports should not be discounted.

Maneuver

Maneuver is the employment of forces on the battlefield through movement in combination with fire and/or time to achieve a position of advantage over the enemy to accomplish the mission. However, gaining positional advantage may be inconsequential when not quickly exploited by violent combat. A tough and dedicated opponent will seldom capitulate when placed in an untenable position. The enemy will accept the cost of maneuvering to another position under indirect fires to avoid decisive combat on our terms. Once an advantage is gained through maneuver, forces exploit that maneuver with close combat to obtain a decisive victory.

Maneuver requires mental and physical agility. Mentally, commanders must be able to visualize the operation, determine the critical events, and develop a scheme of maneuver that will be successful. Physically, maneuver requires the requisite mobility means to enable the GCE's employment against the enemy at the time and place of our choosing. This physical and mental agility is created by thorough training, preparation, appropriate task organization, relevant doctrine, mission tactics, and reliable equipment.

The commander should avoid decisive combat prior to discovery of an enemy critical vulnerablility. In this respect, the GCE cannot become so embroiled in combat with ancillary enemy forces that the bulk of its combat power is not easily maneuvered. When the time to maneuver decisively against an enemy critical vulnerability is possible, the GCE must respond with all of its strength at the decisive time and place. The GCE commander must understand the purpose his force has within the MAGTF commander's plan.

Main Effort

During any phase of an operation, *any element* of the MAGTF may be designated the main effort to accomplish a critical task. The CSSE may be designated the main effort during a period requiring significant sustainment operations. The ACE may be designated the main effort during a phase requiring a heavy reliance on aviation. However, organizations designated the main effort before or during ground combat are so designated to facilitate decisive maneuver by the GCE. At the decisive time and place, the GCE is designated the main effort and employed to achieve the decision.

Enemy Situation

On an uncertain battlefield, the commander attempts to make contact with the enemy with the smallest possible friendly force. This contact will usually be made between the enemy and friendly reconnaissance units or other units tasked with probing enemy dispositions. Based on the information produced by initial and subsequent contacts, the commander constantly develops the situation, determines the enemy vulnerability, and then strikes with the bulk of his combat power. The bulk of the GCE commander's combat power is uncommitted and protected from enemy fires. These forces provide the commander the flexibility to rapidly conduct decisive maneuver. When the situation is well known to the commander, in terms of his own force and that of the enemy, overwhelming combat power may be applied simultaneously throughout the depth of the enemy force. However, in either case, the GCE commander's scheme of maneuver must be flexible enough to adjust rapidly to changes in the situation, as even the products of the best intelligence change once the battle has been joined.

Mobility

Mobility is a battle enhancer. Mobility of the force provides the means to move faster than the enemy and concentrate against him. This applies equally to fire support assets that are resident in the GCE or support it. It is a fundamental aspect of speed, which allows the GCE to catch the enemy off guard. Mobility of the GCE must be greater than or equal to that of the enemy. Wheeled and tracked vehicles provide capability to move rapidly; movement by foot provides the ability to deny the enemy exclusive rights to rugged terrain; and the ACE provides a means of quickly establishing a ground force at practically any location on the battlefield and bringing to bear significant firepower on the enemy. The integration of mobility means by the GCE commander in his scheme of maneuver is determined by—

- Enemy capabilities.
- Timing of critical events.
- Maneuver space.
- Friendly capabilities.

Mass

The ability to concentrate and move quickly provides the commander the ability to mass. Mass relates to combat power, mobility, and speed and must be applied in such a manner as to successfully attack a critical vulnerability. Mass cannot be confused with volume of fire or personnel and equipment density alone. It is more important to maximize the mobility of the GCE by speed than the actual numbers of units moved at the same speed. For example, it is better to provide all available mobility assets to one maneuver unit or organization and to use that unit to strike the enemy unexpectedly than to distribute mobility assets throughout the force and move at a uniformly slower rate.

Sustainment

Mobility of the GCE is enhanced by properly planned MAGTF sustainment. An adequate and responsive MAGTF sustainment capability legitimizes the concept of operations. The CSSE must be responsive to the GCE in a rapid environment over an extended battlefield. The GCE commander must weight his main effort with sustainment support (both internal to the GCE and external from the CSSE) as well as with maneuver units and firepower.

Firepower

Firepower is the amount of fire that may be delivered by a position, unit, or weapon system. Direct and indirect fires are the means by which the GCE kills the enemy and destroys his equipment. Firepower is resident at all echelons of the GCE. To maximize speed and maintain momentum, the GCE should apply the most responsive firepower assets first to overcome enemy resistance. Usually, the most readily available means are those organic to the GCE since these are available 24 hours a day regardless of weather conditions. When these means are inadequate or not the most effective, then assets are obtained from outside the GCE. However, the application of fires from other agencies may require additional time and coordination.

To maximize the benefits of the firepower component of maneuver, target acquisition capabilities must provide near real-time detection and recognition. Assessment of target damage must be rapid and accurate. Only on the basis of accurate battle damage assessment (BDA) from multiple sources does the commander launch his maneuver force. Organic target acquisition capabilities of the GCE are relatively limited to ground radar/sensor detection and visual observation. As such, the GCE's reliance on and connectivity to the CE and ACE reconnaissance, surveillance, and target acquisition (RSTA) capabilities cannot be overstated. Historically, target acquisition by the GCE occurs after contact, creating a reactive vice proactive maneuver or firepower response.

GCE commanders must ensure that R&S plans facilitate the employment of available weapon systems throughout the depth of the enemy's formation. Gaps in the employment of the GCE RSTA effort are addressed by the GCE commander through coordination and integration of MAGTF and naval expeditionary force (NEF) assets. Identification of critical information requirements and high-payoff targets require thorough planning to ensure efficient employment of collection resources.

Effects of Firepower

An understanding of firepower begins with an understanding of the effects it can produce on the enemy. We use firepower to create one of four effects—destruction of enemy personnel and equipment; neutralization of targets; suppression of the enemy when in direct contact; and harassment to disturb the rest of the enemy troops, to curtail movement, and to lower morale. To accomplish these effects, the GCE contains a variety of weapon systems that enhance its operations and contribute to the MAGTF as a whole.

Artillery

Artillery is the primary firepower asset in the GCE. Artillery conducts three tasks. Artillery—

- Provides timely, close, accurate, and continuous fire support.
- Provides depth to combat by attacking hostile reserves, restricting movement, providing long-range support for reconnaissance elements, and disrupting C^2 systems and logistical installations.

- Delivers counterfire within the range of the weapons systems to ensure the freedom of action of the ground forces.

Infantry Mortars

Infantry mortars provide responsive, close, and continuous fire support to infantry and light-armored reconnaissance battalions and below. They are ideal for attacking close-in targets, targets on reverse slopes, urban targets, and targets in areas difficult to reach with low-angle fire, and suppression of immediate targets.

Tanks

Tanks are the key element in creating shock effect for the GCE. The firepower, armor protection, and mobility of tanks are well suited for exploiting breakthroughs and conducting counterattacks. They provide precision direct fires against enemy armor, fighting vehicles, and hardened positions. Combined with aviation forces and surface fire support, tanks and mechanized infantry provide the commander with a potent maneuver force capable of rapidly uncovering terrain and forcing the enemy to fight or displace. The mobility and range of tanks and vehicle-mounted antiarmor guided missile systems allow their employment throughout the battlefield to include security operations.

Fire Support Coordination

Fire support coordination is conducted throughout the GCE. The GCE commander has the facilities to coordinate the fires of his organic assets, those of the ACE in his support, and those of agencies outside the MAGTF in combination against the enemy in support of his scheme of maneuver. Should the fire support coordination facilities of the MAGTF be rendered inoperable, the GCE fire support coordination system is capable of providing real-time coordination for the entire MAGTF. This includes coordination of fires throughout the MAGTF's battlespace.

Terminal fire support controllers within the GCE are capable of controlling naval surface fires, artillery, mortars, and aviation. These controllers provide to the GCE and MAGTF commanders the ability to engage the enemy throughout the depth of the battlefield with a wide variety of ordnance from all available means. Forward air controllers (airborne), provided by the ACE, complement ground controllers and provide additional depth to the fire support effort.

Direct Fire

Direct fire infantry weapons are most effective when combined and coordinated with indirect fires and effective maneuver. The GCE's close, direct fire battle is not the commander's coup de grace. It is an integral part of the scheme of maneuver and requires the same degree of planning and forethought that ordinarily exists in fire support planning.

For all that the GCE brings to the fight in support of its own assigned tasks and the MAGTF as a whole, the GCE relies on the MAGTF, principally the ACE, to engage the enemy at long range with an extended array of weapons and ordnance. The ability of the MAGTF commander to tailor the appropriate fire support means at the farthest reaches of the MAGTF's battlespace is critical to the GCE's freedom of action and employment at the decisive time and place.

Security

The GCE provides to the MAGTF a number of means for security operations. Security forces may consist of task organized light-armored reconnaissance units, tanks, mechanized infantry, and artillery. These forces are tailored for screening, guarding, or covering the force and can shift quickly to accomplish other missions once their security mission is completed. The GCE may be tasked to provide response forces in the rear area; provide security for critical installations when not the main effort; conduct security operations at critical areas on the MAGTF's flank; or ensure physical connectivity with adjacent units. However, the GCE must not be so burdened with responsibility for MAGTF security that it is not capable of conducting decisive maneuver at the necessary time and place.

Chapter 3

Command and Control

Command and control (C^2) is the exercise of authority and direction by a properly designated commander over assigned forces in the accomplishment of the mission. C^2 is influenced by the internal requirements of the GCE and the battlespace that can be dominated. Regardless of the size of the organization, all commanders must share a common perspective of the battlespace and have the ability to acquire critical information. The commander and his subordinate commanders must see the battlespace in a similar manner to exploit the full potential of the MAGTF and GCE firepower, maneuver, and sustainment capabilities. At every echelon of command, C^2 has to be flexible, fast, and decentralized. The commander provides the impetus for effective C^2.

The Commander

The commander influences the unit under his command by his personality, attitude, technical and tactical proficiency, and leadership. Resoluteness is an essential trait in the commander. A commander cannot plead absence of orders as an excuse for inactivity. Commanders who merely wait for orders cannot exploit the opportunity of the moment.

The commander discharges his responsibilities by sound planning, timely decisions, clear definitive orders, personal supervision, and exemplary leadership. The commander must have the capacity to withstand the fluctuations and physical and emotional shocks of combat without loss of effectiveness. In spite of the most careful planning and anticipation, unexpected obstacles and mistakes are common occurrences in battle. A commander must train himself to regard these events as commonplace and not permit them to frustrate him in the accomplishment of the mission.

When a mission is received, the GCE commander conducts a METT-T-based mission analysis, the products of which he uses to guide the execution of the command function. The GCE commander considers the MAGTF commander's intent and the intent of the MAGTF commander's superior; then the GCE commander determines the end state he must produce to accomplish the mission. The end state with respect to the GCE is then reflected in the GCE commander's intent. The GCE commander assigns his subordinates missions and task organizes his force accordingly to achieve the desired end state.

The commander's intent provides a road map to achieving conditions necessary for decisive maneuver. It allows subordinates to make decisions in a fluid environment in the absence of orders. It is a clear, concise statement that defines success for the force as a whole by establishing, in advance of events, the general conditions he wants to obtain at the conclusion of the battle or operation. It may provide guidance from the commander as to where and how risk will be accepted. Once articulated and disseminated, it stimulates the entire planning process, unifies the force toward a common mission objective, and provides subordinate commanders with a focus on which to gauge freedom of action. The commander's intent is not a summary of the concept of operations. Commander's intent—

- Expresses the purpose of the operation.
- Describes critical vulnerabilities and center of gravity for both friendly and enemy forces.
- Provides a vision of how the operation will be conducted in a broad scope.
- Describes the desired end state as it relates to the enemy, friendly forces, terrain, and future operations.

Command and Control Organization

The GCE C^2 organization provides for control of maneuver, fire support, intelligence operations, aviation support, CSS, and C^2W. To accomplish these control tasks, the headquarters of the GCE is organized into three echelons—tactical, main, and rear.

Tactical Echelon

The tactical echelon provides the commander freedom of movement and the information required to maintain situational awareness. Its primary function is to place the commander at his main effort. The tactical echelon ordinarily consists of representatives from the G-2/S-2 and G-3/S-3 (to include fire support and aviation). During operations, the GCE commander normally moves forward to personally observe and influence the course of the battle. The GCE commander may move to observe the progress of units during critical events, such as passage of lines, battle handover, and attacks against critical vulnerabilities, or to conduct a personal reconnaissance. The tactical echelon supports the commander's ability to move about the battlefield.

Main Echelon

The primary interests of the main echelon are directing current operations and planning future operations. Depending on the level of command, the main echelon may be divided into two sections—one handling current operations and the other planning future operations.

Current Operations Section

The current operations section supervises and coordinates ongoing combat operations. This includes the combat, combat support, and combat service support necessary to prosecute the current battle.

Future Operations Section

The future operations section monitors current operations and plans future operations. Deep operations, as defined by the GCE commander in terms of space, time, and threat, are planned by the future operations section. The future operations section coordinates the contingency plans and the necessary actions of all subordinate organizations and supporting units in the execution of future operations.

Rear Echelon

The principal function of the rear echelon is to support combat operations by providing C^2 of rear area operations. Tasks that are supervised throughout the rear area include rear area security operations, terrain management, sustainment, movement control, and associated functions. The rear echelon must be capable of monitoring the activities of the forward units and the other two headquarters echelons.

The Command Post

Any one of the three echelons may function as the command post. At any one time, there is only one command post which is dictated by the commander's physical location.

Command and Control Support

Command and control support is the planned complementary employment of all information-related systems, assets, and associated resources so that the flow and processing of information is deliberately controlled to advantage in support of the GCE commander's decision and execution cycle. In this regard, the commander expands his ability to reason with the capabilities of automated systems to transport, manipulate, fuse, store, and recall information; the capabilities of his collection assets to provide information; and the capabilities of his staff to identify that information critical to the tactical situation and to determine what that critical information conveys.

C^2 support forces include the personnel, equipment, facilities, and communications that provide reconnaissance, tactical air control, electronic warfare, fire support coordination, automated data processing, sensor management, signals intelligence, space systems, deception, and other information-related services. C^2 support helps create a common situational awareness that speeds the ability of the commander and key personnel to convey and share ideas quickly to enhance unity of effort and tempo of operations.

The Staff

Command authority is extended through control to link decision-making to execution. The GCE commander relies on his staff to control and coordinate planning and execution of the operation within the guidance of his intent. The commander limits the size of his staff to the minimum personnel needed to function effectively. The staff develops and implements the commander's plan and serves the commander in order to allow him the freedom to exercise command. The staff also supports subordinate commanders to ensure the successful accomplishment of assigned tasks.

The staff implements the function of control. *Control* is that authority exercised by a commander over part of the activities of subordinate organizations or other organizations not normally under his command, which encompasses the responsibility for implementing orders or directives. The staff must anticipate requirements for action and planning. It must be proactive and prepared to provide the commander the estimate of the current situation, courses of action to meet tactical or operational needs, and the capability to dictate the tempo of decisionmaking/execution/feedback as the time to plan and execute is limited. The staff should have the authority to act in the name of the commander in order to execute the commander's plan. As the staff is the conduit for information and must manage information flow throughout the command, organization of the tactical, main, and rear headquarters echelons must facilitate information management.

Commander's Preparation of the Battlespace

The commander's preparation of the battlespace occurs at each level of command. This preparation arms each commander with critical knowledge of his own force, the enemy, time, and space so that the commander can prepare the battlespace according to his desires. It further enhances the commander's visualization of the battlespace to determine how the commander might task organize and position his force during different phases of the operation. The commander's preparation provides answers to the commander's critical information requirements, prepares the battlespace for shaping, permits effective maneuver and fires, and disrupts the enemy plan and disperses his forces.

The battlespace is prepared through surveillance, reconnaissance and counter-reconnaissance, intelligence preparation, and targeting. The higher the level of command, the more formal the process. The MAGTF commander's preparations involve all elements of the MAGTF. The GCE commander must utilize his resources to meet the demands of the MAGTF commander and his own preparation requirements.

The commander must organize the time available for his own planning and preparation and the planning and preparations of his subordinate commanders. In addition, the GCE commander must organize his available time to meet any external timelines as directed by the MAGTF commander. In particular, a rehearsal of the plan should always be conducted when the time permits. When conducting a rehearsal, the commander should emphasize key events that trigger friendly actions. The rehearsal is a tool the commander uses to reinforce understanding of the plan and to help subordinate commanders visualize the commander's intent and what they can do when the battle does not go according to plan. Oral orders supported by rehearsals have more value than written orders without rehearsals.

IPB is an integral component of the commander's preparation of the battlespace. It is a systematic and continuous approach to analyzing the enemy, weather, and terrain in the specific geographic location.

IPB integrates enemy doctrine with information on the terrain and weather as they relate to the mission and specific battlefield conditions. It assists the commander in evaluating enemy capabilities, vulnerabilities, and probable courses of action. The commander, operations officer, intelligence officer, fire support coordinator, and other staff members have critical roles and responsibilities to complete the IPB process. Decisions regarding the employment of intelligence collection assets, identification of high-payoff targets, assignment of fire support systems, development of the scheme of maneuver, and the placement or allocation of CSS are products of comprehensive staff participation.

Organization of the Battlefield

During the commander's preparation, the battlefield is organized to establish the desired relationship between the subordinate elements of the friendly force and the enemy in time, space, and function. The commander must be able to see the entire battlefield and shape or exploit each of its dimensions to his tactical advantage. No dimension can be ignored, and the commander cannot consider any dimension in isolation. The higher the level of command, the larger and more complex the battlefield becomes and the broader the commander's field of vision and appreciation must be. The battlefield is organized into three roughly concentric areas—the area of operations, the area of influence, and the area of interest. Furthermore, from the commander's perspective, three related operations are conducted throughout the battlespace—deep, close, and rear operations as further discussed under Battlespace Operations on page 3-11.

Area of Operations

The area of operations is a portion of an area of war necessary for military operations and the administration of such operations. The area of operations may be a zone of action in the offense, a sector in the defense or retrograde, or a tactical area of responsibility (TAOR). When the commander assigns an area of operation to a subordinate commander, the following should be considered:

- Higher unit capabilities.
- Subordinate unit capabilities.
- Adjacent unit capabilities.
- Unprotected flanks.
- Close, deep, and rear operations of higher and subordinate commanders.

Area of Influence

The area of influence is a geographical area wherein a commander is directly capable of influencing operations by maneuver or fires normally under his command or control. The area of influence is not an assigned area; rather, it is based on friendly capabilities. The area of influence normally extends to the limits of supporting arms controlled by the unit. For the GCE, this usually is interpreted as the effective range of artillery, and, for the MAGTF, it is the range of organic aviation. A higher commander should not assign a subordinate commander an area of operations beyond the subordinate commander's influence.

Area of Interest

The area of interest is that area of concern to the commander, in-cluding the area of influence and areas adjacent to it and extends to the objectives of current or planned operations. This area also in-cludes areas occupied by enemy forces who could jeopardize the ac-complishment of the mission. The size of the area of interest depends on the situation and is based solely on the commander's concerns. At successively higher levels of command, the com-mander must plan further into the future; consequently, the area of interest will be larger. Likewise, a mobile enemy force may dictate that the area of interest be large due to the enemy's ability to move great distances in a short time.

Figure 3-1. Organization of the Battlefield.

Airspace

The friendly and enemy use and control of airspace is an essential dimension of ground combat which the commander must always consider. Aviation units provide reconnaissance, fire support, and logistic support to the ground battle. They can facilitate C^2. Aviation units also permit rapid maneuver and massing of forces over otherwise impassable terrain. They allow the establishment of air lines of communication where ground lines of communication do not exist. When establishing security for his force, the commander must ensure security against observation and attack from the air as well as from the ground. For senior commanders, this includes security against satellite imagery and other remote intelligence-gathering assets.

Time

As he executes the current operation, the commander must shape the battlefield for anticipated future operations by establishing conditions that will give him the tactical advantage when the time comes to execute those operations. Anticipating the course of the battle and preparing for it are essential to maintaining the initiative. The commander who fights only in the present will invariably be reacting to the enemy and fighting in accordance with the enemy's will.

The higher the level of command, the longer the time needed to develop, disseminate, and execute a plan, and, consequently, the further ahead in time the commander must be thinking. The period into the future the commander must plan depends on the situation, especially the enemy's mobility and long-range firepower.

Time-Distance Appreciation

Seeing the battlespace in terms of time requires an appreciation for time-distance factors. Time-distance appreciation is critical to establishing and dominating the tempo of operations. Based on the situation, the GCE commander must continually and realistically calculate the—

- Time necessary to complete movements, maneuvers, preparations, or other actions.
- Time before the force can expect to close with the enemy.
- Distance from the main body that security forces must operate to provide ample protection.
- Enemy capabilities regarding time-distance to move its forces.
- Amount of delay that can be imposed on the enemy through interdiction or other means.
- Other considerations that affect the plan.

Battlespace Operations

The GCE commander plans, organizes forces to support, and conducts deep, close, and rear operations throughout the depth of the battlespace to maximize the effects of his resources against the enemy force. These activities are coordinated and executed continuously at all levels of command. The effectiveness of these concurrent operations determines the outcome of the operation.

Deep, close and rear operations are not necessarily characterized by distance or location on the battlefield. Rather, they are functional in nature. Commanders must view the entire battlespace and determine what, where, and when firepower, maneuver, intelligence, and sustainment activities are to be applied against the enemy force.

Deep Operations

Deep operations are military actions conducted against enemy capabilities which pose a potential threat to friendly forces. These military actions are designed to isolate, shape, and dominate the battlespace and to influence future operations. Deep operations are conducted primarily through the employment of fires. They seek to open the window of opportunity for decisive maneuver and are designed to restrict the enemy's freedom of action, disrupt the coherence and tempo of his operations, nullify his firepower, disrupt his C^2, interdict his supplies, isolate or destroy his main forces, and break his morale.

The enemy is most easily defeated by fighting him close and deep simultaneously. Well-orchestrated deep operations, integrated with simultaneous close operations, may be executed with the goal of defeating the enemy outright or setting the conditions for successful future close operations. Deep operations enable friendly forces to choose the time, place, and method for close operations.

Deep operations in the MAGTF are primarily planned, coordinated, and executed by the MAGTF CE. Although deep operations are primarily the responsibility of the MAGTF CE and may be conducted largely with ACE resources, the GCE has a significant role. The GCE contributes to the deep operations of the MAGTF by recommending deep operations objectives and targets that will help shape the future GCE battlespace. The GCE must also be prepared to provide resources to execute deep operations as directed by the MAGTF CE and may, in fact, be tasked to control certain deep operations missions on behalf of the MAGTF. Additionally, the GCE may plan and execute deep operations within its own area of operations to shape the GCE battlespace.

Deep operations may include—

- Deception.

- Deep interdiction through deep fires, deep maneuver, and deep air support.
- Deep surveillance and target acquisition.
- Command and control warfare.
- Offensive antiair warfare.

Deception plays a major part in shaping the battlespace, and the GCE plays a major role in the MAGTF's deception operations. Deceptive measures such as demonstrations or feints can disrupt enemy plans, divert enemy forces away from the actual point of battle, and delay enemy reactions thereby placing the enemy at a disadvantage when forces come in contact. Deception activities of the GCE must be well coordinated with the MAGTF CE to ensure efforts are directed towards a common goal. GCE deep interdiction capabilities may include the long-range fires of its artillery and rockets and its high-speed, mobile maneuver forces. The GCE may contribute to deep surveillance and target acquisition efforts with its organic reconnaissance forces and the counterbattery radar assets of artillery units. Command and control warfare and offensive antiair warfare conducted in support of deep operations are largely outside the capability of the GCE.

The coordination and integration of MAGTF and GCE deep operations help to ensure constant pressure on critical enemy capabilities throughout the battle. Because of the scarcity of resources with which to conduct these activities, deep operations must be focused on those enemy capabilities that most directly threaten the success of the projected friendly operations.

Close Operations

Close operations are military actions conducted to project power decisively against enemy forces which pose an immediate or near-term threat to the success of current battles and engagements. These military actions are conducted by committed forces and their readily available tactical reserves, using maneuver and combined arms. These operations require speed and mobility to enable the rapid concentration of overwhelming combat power at the critical time, application of that combat power at the critical place, and the ruthless exploitation of success gained. The opportunity to achieve a decision will be lost if a commander fails to exploit success.

Rear Operations

Rear operations are those actions necessary to sustain deep and close operations. Rear operations ensure the freedom of action of the force and the ability to conduct continuous operations. Tasks associated with rear operations include CSS, terrain management, and security. As these activities are conducted throughout the area of operations, they should not be considered solely by geographic location. The commander will, in fact, conduct rear operations throughout the battlespace to support the conduct of the single battle.

Chapter 4

Operational Maneuver
From the Sea

Operational maneuver from the sea (OMFTS) is the application of maneuver warfare to operations in a maritime environment. Applying the principles of maneuver warfare to expeditionary operations, OMFTS can exploit the extraordinary operational mobility offered by naval expeditionary forces without loss of momentum by conducting a seamless operation from the sea to the objective. Seamless operations prevent loss of initiative during transition from one phase to another which is key to controlling tempo. Tempo is used as a weapon to create conditions that can paralyze the adversary's decisionmaking capability.

OMFTS enables Marine forces to take maximum advantage of their capabilities in order to disrupt the cohesion of the enemy while avoiding attacks from enemy strength. OMFTS directs combat power toward a critical vulnerability to blind, confuse, and defeat the enemy through simultaneous use of sea, air, space, and land forces.

Execution of OMFTS

OMFTS requires a level of execution that can be conducted only by a standing integrated naval staff that understands operational art and maneuver warfare. This staff must understand the unique requirements of OMFTS and the capabilities that OMFTS provides the overall commander of the operation or campaign. OMFTS—

Demands Seamless Command and Control

C^2 demanded by OMFTS focuses on planning and executing a continuous, seamless operation from the sea to the distant objective ashore. OMFTS is continuous because it elevates the principle of *unity of command* to a constant common denominator—regardless of operational phases, geographic divisions, or battlespace expansion. OMFTS is seamless because its twin pillars of staff integration and superior operational tempo do not support any disruption of operational continuity.

The definition of a successful operation is unity of effort pointed to the common goal—the operational objective. The NEF commander expresses his operational intent for the course of the campaign. His intent is a visualization of how the force will achieve the desired end state—from planning to mission accomplishment. The C^2 of the MAGTF is inextricably tied to the C^2 of the NEF to permit OMFTS without interruption from the sea base to the objective. This eliminates the difficulties of moving C^2 ashore.

Attacks Critical Vulnerabilities

OMFTS is directed towards operational objectives in support of the strategic aim. Instead of focusing on the seizure of terrain, OMFTS applies combat power directly to critical vulnerabilities. OMFTS does not envision the methodical buildup of combat power ashore. This momentum, in turn, threatens the enemy's decision cycle to react to our attack. The NEF uses the entire array of combined arms to shape the battlespace and initiate decisive maneuver against objectives that are critical to the enemy's ability to mass, maneuver, and control his force.

Capitalizes on Unpredictability

OMFTS limits the ability of the enemy to predict NEF operations. The NEF can conduct operations from over-the-horizon to achieve, at a minimum, tactical surprise, and it exploits that surprise with rapid projection of power against inland objectives. Unlike linear amphibious tactics that emphasize strict control of maneuver forces during ship-to-shore *movement*, OMFTS emphasizes ship-to-objective *maneuver*. The NEF conceals its operational intent by—

- Using integrated national, theater, and organic intelligence capabilities that identify gaps in the enemy defensive system.

- Conducting offensive C^2W and advance force operations.

- Applying surface, subsurface, aviation, and joint/combined fires that can be applied throughout the battlespace to mask the main effort.

- Taking advantage of emerging enhanced technologies to increase our range of force projection options to overcome previously impassable terrain.

Uses the Sea as Maneuver Space

The sea is our maneuver space to disperse the force for offensive and defensive operations. Traditional amphibious forces habitually mass forces prior to execution, clearly signaling intent. OMFTS disguises the main effort and masks the axis of advance by initiating decisive maneuver from dispersed locations. Attacks launched from dispersed locations—

- Complicate enemy targeting efforts.
- Enhance deception.
- Expand the littoral battlespace.

Maximizes Seabased Logistics

Sustainment is key to maintaining the momentum of the attack. Rather than rely on the systematic buildup of sustainment ashore, tailored logistic packages are "pulled" by or "pushed" to the maneuver units as the situation dictates. This requires anticipatory planning to ensure continuous support as forces maneuver. Seabased logistics—

- Increase the survivability of logistic resources.
- Focus on critical sustainment needs of the maneuver force.
- Permit rapid reconstitution of the force afloat.

Amphibious Operations

NEF power projection options range from the use of precision-guided munitions, aviation, special operations forces, and C^2W to the employment of ground forces. Amphibious operations are part of OMFTS and integral to naval power projection. Amphibious operations are conducted within OMFTS to enable the introduction of larger forces and to support a main effort elsewhere or as the main effort in a campaign. The threat of amphibious operations may serve as a deterrent to hostile action; shield intent and objectives; and disperse and fix in place enemy forces over an extended area. Amphibious forces of the NEF conduct one of the following four types of amphibious operations.

Amphibious Assault

An amphibious assault is the principal type of amphibious operation that establishes a force on a hostile shore. Amphibious assaults are essential to the landward dominance of battlespace. Maneuver of the landing force is a logical extension of the maneuver of the amphibious task force. When necessary, an amphibious assault against an integrated defense will require the NEF to focus over-whelming combat power to create a gap. The landing force must then have the C^2, mobility, firepower, and sustainment necessary to exploit this window of opportunity.

Amphibious Raids

An amphibious raid is an attack from the sea involving swift incur-sion into hostile territory for a specified purpose, followed by a planned withdrawal. Raid forces may consist of aviation, infantry, engineers, artillery, or any other element with skills and equipment needed for the mission. Amphibious raids conducted in support of OMFTS are directed against objectives requiring specific effects not possible with other power projection means.

Amphibious Demonstrations

Amphibious demonstrations enhance deception and surprise. A demonstration is conducted to deceive the enemy by a show of force to induce him to adopt an unfavorable course of action. The value of the demonstration must be measured against its merit as a supporting effort and its impact on the main effort. The GCE may provide forces to make the demonstration more plausible. Forces and assets providing an amphibious demonstration can be rapidly redirected to support operations elsewhere.

Amphibious Withdrawals

An amphibious withdrawal is an operation involving the evacuation of land forces by sea in naval ships or craft from a hostile shore. Amphibious withdrawals may be conducted to extract a force under pressure, to assist in the repositioning of forces elsewhere in theater, to reconstitute forces afloat, or to establish an operational reserve after introduction of heavy follow-on forces. Amphibious withdrawals are tactical in nature and therefore more than administrative backloading of amphibious ships.

Organization for Ship-to-Objective Maneuver

OMFTS requires organization of the force for ship-to-objective maneuver. Rather than organizing to support ship-to-shore movement, the force must be organized to permit the seamless and continuous application of combat power to distant inland objectives. The following guidelines apply to organization for ship-to-objective maneuver:

- Provide for the concentration of combat power at the critical time from dispersed locations.
- Provide maximum shock effect at the penetration points to overcome enemy resistance at the beach.
- Provide for the timely employment of combat, combat support, and combat service support elements required to support the commander's concept.
- Provide depth to the assault to ensure exploitation of gaps created or located by the NEF.
- Provide sufficient flexibility to exploit opportunities discovered during execution of the operation.

Chapter 5

Offensive Operations

The decisive form of war is the offensive. The focus of the offensive is the enemy force, not seizure of terrain. Even in the defense, a commander must take every opportunity to seize the initiative by offensive action and to carry the battle to the enemy. Offensive operations are undertaken to—

- Destroy enemy forces and equipment.
- Deceive and divert the enemy.
- Deprive the enemy of resources.
- Gain information.
- Fix the enemy in place.
- Disrupt enemy actions or preparations.

Offensive operations require the attacker to weight the main effort with superior combat power. The requirement to concentrate and the need to have sufficient forces available to exploit success imply accepting risk elsewhere. Local superiority must be created by maneuver, deception, speed, surprise, and economy of force. Success in the offensive is best gained from attacks that—

- Avoid the enemy's main strength; attack him where he is weak.
- Isolate his forces from their sources of supply.
- Force him to fight in an unexpected direction over ground he has not prepared.
- Force the enemy commander to make hasty decisions based on an inaccurate battlefield picture.

Fire superiority is one of the most important requisites in offensive combat. It must be gained early and maintained throughout the attack to permit freedom of maneuver without prohibitive loss. But fire alone can rarely force a favorable decision. The effects of fire must be exploited by maneuver. Fire superiority rests chiefly on supporting arms employed with the organic fires of the attacking units. It depends not only on volume of fire but also on its direction and accuracy and the close coordination of all fires with maneuver.

Types of Offensive Operations

There are four general types of offensive operations—movement to contact, attack, exploitation, and pursuit. Though described in a logical or notional sequence, these operations may occur in any order or simultaneously throughout the battlefield. A movement to contact may be so successful that it immediately leads to an exploitation, or an attack may lead directly to pursuit. Isolated or orchestrated battles will become increasingly rare, as the MAGTF will fight the enemy throughout the depth of the battlespace.

Movement to Contact

Movement to contact is an offensive operation conducted to develop the situation and to establish or regain contact with the enemy. A properly executed movement to contact allows the commander to make initial contact with minimum forces and to expedite the employment and concentration of the force. See figure 5-1. The commander must foresee his actions upon contact. He organizes his force to provide flexible and rapid exploitation of the contact gained. The force utilizes battle drills that focus on overcoming initial contact quickly. These procedures must be practiced and thoroughly rehearsed to permit the entire force to act without detailed guidance. Failure to prepare accordingly results in delay and confusion and grants the enemy time to seize the initiative and to dictate the conditions under which the engagement is fought.

The GCE commander's intent will dictate the extent that his forces will be engaged. The desired contact may be by observation, physical contact between security forces, or physical contact with main forces. In each instance, the task organization, scheme of maneuver, and support required may differ significantly. In addition, the GCE commander must consider and arrange for additional support required from the MAGTF. Every reconnaissance and security means is employed so that the main force will be committed under the most favorable conditions.

Figure 5-1. Movement to Contact.

To maintain his freedom of action once he makes contact, the commander deploys an advance force capable of locating and fixing the enemy. The main body is positioned so as to remain uncommitted, capable of maneuvering without effective enemy interference at the time of the commander's choosing. The advance force must ensure the uninterrupted progress of the main body. The advance force must contain sufficient combat power to overcome security and delaying forces and provide time for the commander to deploy the main body at the critical location. This allows the commander to choose the best possible time and location to exploit the meeting engagement, to maintain pressure on the enemy, and to shift to another type of offensive operation. Premature deployment of the main body is costly in terms of time, resources, and disclosure of the main effort. A movement to contact ends when ground enemy resistance requires the deployment of the main body.

Attack

The purpose of the attack is to defeat, destroy, or neutralize the enemy. An attack emphasizes maximum application of combat power, coupled with bold maneuver, shock effect in the assault, and prompt exploitation of success. There are four principal tasks in an attack:

- Prevent effective enemy maneuver or counteraction.
- Maneuver to gain an advantage.
- Deliver an overwhelming assault to destroy him.
- Exploit advantages gained.

Commanders must expect to make adjustments during an attack. Skillful commanders provide for the means and methods to work these adjustments rapidly in order to maintain the momentum of the attack. Flexibility in the scheme of maneuver, organization for combat, and universal understanding of the commander's intent provide means to adapt to these changes on the battlefield.

The commander presses the attack although his troops may be exhausted and his supplies depleted against a weakened or shaken enemy. However, the commander must strive to accomplish his objectives prior to the force reaching its culminating point, that point in time or location that the attacker's combat power no longer exceeds that of the defender. Once reaching his culminating point, the GCE commander risks overextension and counterattack by the defender who recognizes these vulnerabilities.

There are no concrete criteria for determining when an organization involved in an engagement, battle, or campaign has reached its culminating point. Rather, it must be an intuitive understanding borne of experience to which the commander must be ever sensitive. If the force is incapable of accomplishing its mission before reaching its culminating point, the commander must plan to phase his operation accordingly. The differences between the types of attacks lie in the degrees of preparation, planning, coordination, and the effect desired on the enemy.

Hasty Attack

A hasty attack is an attack in which preparation time is traded for speed to exploit opportunity. To maintain momentum or retain initiative, minimum time is devoted to preparation. Those forces readily available are committed immediately to the attack. A hasty attack seeks to take advantage of the enemy's lack of readiness and involves boldness, surprise, and speed to achieve success before the enemy has had time to improve his defensive posture. By necessity, hasty attacks are simple and require a minimum of coordination with higher and adjacent commanders. Hasty attacks are most likely the result of movements to contact, meeting engagements, penetrations, or fleeting opportunities created by disorder, enemy mistakes, or the result of our own actions.

To minimize the risks associated with the lack of preparation time, organizations should utilize standard formations and proven standing operating procedures and conduct rehearsals. Major reorganization of the force should be avoided and habitual relationships maximized when the commander task organizes the force.

Deliberate Attack

A deliberate attack is a type of offensive action characterized by preplanned coordinated employment of firepower and maneuver to close with and destroy the enemy. The deliberate attack is a fully coordinated operation that is usually reserved for those situations where the enemy defense cannot be overcome by a hasty attack or where the deployment of the enemy shows no identifiable exposed flank or physical weakness. Deliberate attacks usually include a high volume of planned fire, deception plans, extensive use of C^2W, and all-source intelligence gathering. Time taken by the commander to prepare a deliberate attack is also time in which the enemy can continue defensive improvements, can disengage, or can launch a spoiling attack.

Spoiling Attack

Commanders normally mount spoiling attacks from a defensive posture to disrupt an expected enemy attack. A spoiling attack attempts to strike the enemy while he is most vulnerable—during his preparations for attack in assembly areas and attack positions or while he is on the move prior to crossing the line of departure. Spoiling attacks are conducted similarly to any other type of attack. Frequently, the circumstances in which commanders conduct spoiling attacks preclude full exploitation. However, when the situation permits, a spoiling attack should be exploited, and commanders must be prepared to take advantage of the success like that achieved in any other attack.

Counterattack

Commanders conduct counterattacks either with a reserve or otherwise uncommitted or lightly engaged forces. The counterattack is conducted after the enemy has commenced his attack and a resolute defense or enemy tactical error exposes him to effective counteraction. See figure 5-2.

The commander must identify the enemy's main effort and determine the appropriate course of action to prevent it from succeeding. Ordinarily, the commander will develop a number of counterattack

Figure 5-2. Counterattack.

options to expedite the execution of the counterattack. The commander plans engagement areas throughout the defense that permit fires to blunt any penetration by the enemy main effort. Once the enemy main effort is identified and the penetration has been halted or momentum of the attack slowed, the counterattack is launched against the enemy's flank or rear.

Timing of the counterattack and the effort of supporting units is a difficult undertaking and each option requires thorough rehearsal. Counterattacks may also take on the characteristics of a hasty attack. However, these become more difficult to exploit and frequently require the counterattack force to revert to a defensive posture, rather than achieving full exploitation, pursuit, or resumption of the offense.

Feint

A feint is a supporting effort designed to divert or distract the enemy's attention away from the main effort and involves physical contact with the enemy. A feint must be sufficiently strong to confuse the enemy as to the location of the main effort. Ideally, a feint causes the enemy to shift forces to the diversion and away from the main effort. Feints are usually shallow, limited-objective attacks conducted before or during the attack of the main effort. A unit conducting a feint usually attacks on a wider front than normal, with a consequent reduction in mass and depth. A unit conducting a feint normally keeps only a minimal reserve to deal with unexpected developments.

Demonstrations are related operations, also designed to divert enemy attention to allow decisive action elsewhere. A demonstration is a show of force that threatens an attack at another location but does not make contact with the enemy. The commander executes a demonstration by an actual or simulated massing of combat power, troops movements, or some other activity designed to indicate the preparations for or beginning of an attack at a point other than the main effort.

Reconnaissance in Force

The reconnaissance in force is always a deliberate attack by major forces to obtain information and to locate and test enemy dispositions, strengths, and reactions. While the primary purpose of a reconnaissance in force is to gain information, the commander must be prepared to exploit opportunity. A reconnaissance in force usually develops information more rapidly and in more detail than other reconnaissance methods. If the commander must develop the enemy situation along a broad front, the reconnaissance in force may consist of strong probing actions to determine the enemy situation at selected points. See figure 5-3.

The commander may conduct reconnaissance in force as a means of keeping pressure on the defender by seizing key terrain and uncovering enemy weaknesses. The reconnoitering force must be of a size and strength to cause the enemy to react strongly enough to disclose his locations, dispositions, strength, planned fires, and planned use of the reserve. Since a reconnaissance in force is conducted when knowledge of the enemy is vague, a well-balanced force normally is used. Deciding whether to reconnoiter in force, the commander considers—

- His present information on the enemy and the importance of additional information.
- Efficiency and speed of other intelligence collection assets.
- The extent to which his future plans may be divulged by the reconnaissance in force.
- The possibility that the reconnaissance in force may lead to a decisive engagement that the commander does not desire.

Figure 5-3. Reconnaissance in Force.

Raid

A raid is an offensive operation, usually small-scale, involving a penetration of hostile territory for a specific purpose other than seizing and holding terrain. It ends with a planned withdrawal upon completion of the assigned mission. The organization and the composition of the raid force are tailored to the mission. Raids are characterized by surprise and swift, precise, and bold action. Raids are typically conducted to—

- Destroy enemy installations and facilities.
- Capture or free prisoners.
- Disrupt enemy C^2 or support activities.
- Divert enemy attention.
- Secure information.

Raids may be conducted in the defense as spoiling attacks to disrupt the enemy's preparations for attack; during delaying operations to further delay or disrupt the enemy; or in conjunction with other offensive operations to confuse the enemy, divert his attention, or disrupt his operations. Raids require detailed planning, preparation, and special training. Raids conducted with other operations are normally controlled by the local commander.

Exploitation

The enemy may still be capable of fielding cohesive units after being attacked. In the exploitation, the attacker extends the destruction of the defending force by maintaining constant offensive pressure. The objective of the exploitation is the disintegration of enemy forces to the point where he has no alternative but surrender

or flight. When an attack succeeds, the enemy may attempt to disengage, withdraw, and establish or reconstitute an effective defense.

Rapid exploitation of successful attacks inhibits the enemy's ability to do so. Attacks that result in annihilation of the defending force are rare.

The commander must be prepared to exploit every attack without delay. While exploitation following an attack is fundamental, it is especially important in a deliberate attack where the concentration necessary for success requires accepting risk elsewhere. Failure to exploit aggressively the success of the main effort may provide the enemy sufficient time to detect and exploit those risks. As a result, the enemy regains both the initiative and the advantage.

The GCE commander's principal tool for the conduct of an exploitation is his reserve. He may also designate other exploiting forces through the issuance of a fragmentary order. Commanders of exploitation forces must be given as much freedom of action as possible, and efforts must be characterized by boldness, aggressiveness, and speed. However, the commander needs sufficient centralized control to concentrate his forces and to prevent his units from becoming overextended.

Essential to the exploitation is the knowledge of the enemy's condition and identification of enemy critical vulnerabilities. The GCE commander's knowledge of the situation must be so complete that he will not commit his exploitation force prematurely or lose an opportunity by acting too late. Events, such as increased enemy prisoners of war (EPWs), lack of organized defense, loss of enemy unit cohesion upon contact, and capture of enemy leaders indicate an opportunity to shift to an exploitation. Once begun, an exploitation is executed relentlessly to deny the enemy any respite from pressure. Typical missions for the exploitation force include cutting lines of communication, isolating and destroying enemy units, and disrupting enemy C^2.

Pursuit

When it becomes clear that organized enemy resistance has completely broken down, the commander shifts to the pursuit. The difference between an exploitation and a pursuit is the condition of the enemy. The object of a pursuit is annihilation of the enemy force. Like exploitation, pursuit requires broad decentralized control and rapid movement.

The GCE commander must ensure that all assets, to include allocated MAGTF assets, are used to maximum effectiveness during the pursuit. The commander task organizes the force into a direct pressure force and an encircling force. See figure 5-4.

The direct pressure force must have sufficient combat power to maintain pressure on the enemy. The encircling force must have continuous fire support and greater mobility than the enemy. The capabilities of the ACE make it particularly valuable as an encircling force by destroying and denying the enemy routes of escape. To maintain tempo and pressure, the MAGTF commander may shift the main effort to the ACE during a pursuit. A pursuit is pushed to the utmost limits of endurance of troops, equipment, and especially supplies. If the force must stop for rest, maintenance, or reorganization, the enemy may be able to pull together scattered units, emplace obstacles, or break contact altogether.

Figure 5-4.　Pursuit.

Forms of Maneuver

The GCE commander selects the most decisive form of maneuver to achieve his purpose. The forms of maneuver are the frontal attack, flanking attack, envelopment, and the turning movement. While frequently used in combination, each form of maneuver attacks the enemy in a different way and poses different opportunities and challenges to the GCE commander.

Frontal Attack

A frontal attack normally involves attacking the enemy on a broad front by the most direct route. Frontal attacks are used when the attacker possesses overwhelming combat power against disorganized forces or lightly held positions. Frontal attacks are conducted as rapidly as possible to deny the enemy time to react and to sustain the attacker's momentum. See figure 5-5.

The frontal attack is most often selected by commanders tasked with conducting attacks in support of the main effort, during a pursuit, or for fixing an enemy in place. The goal of a frontal attack is to achieve a penetration. As forces attack frontally, successful units
will rupture portions of the enemy defense. The GCE commander may conduct feints or demonstrations in other areas to weaken the enemy effort at the breach by causing him to shift reserves to our advantage. The main effort may be shifted and combat power is brought to bear at the point of penetration to widen the breach, to defeat enemy counterattacks, and to attack the enemy in depth. Successful penetrations—

- Disrupt enemy C^2.
- Force the enemy to expend resources against supporting attacks.
- Force the premature commitment of the enemy reserve.
- Create psychological paralysis in enemy commanders.

Figure 5-5. Frontal Attack.

Flanking Attack

A flanking attack is a form of maneuver where the main effort is directed at the flank of an enemy. A flank may be created by fires, terrain, and enemy dispositions. A flanking attack seeks to strike the enemy's main force while avoiding the frontal orientation of main weapon systems. A flanking attack is similar to an envelopment but is conducted on a shallower axis and is usually less decisive and less risky than a deeper attack.

Figure 5-6. Flanking Attack.

Envelopment

An envelopment is an offensive maneuver in which the main effort passes around or over the enemy's principal defensive positions to attack the objective while avoiding the enemy's main combat power. By nature, it requires surprise, superior mobility (ground and/or air), and successful supporting efforts. An envelopment generally—

- Strikes the enemy where he is weakest.
- Severs enemy lines of communication.
- Disrupts enemy C^2.
- Interrupts enemy CSS.
- Forces the enemy to fight on a reverse front.
- Minimizes the attacker's losses.
- Compels the defender to fight on ground of the attacker's choosing.

Figure 5-7. Envelopment.

The enveloping force avoids the enemy's strength en route to the objective. Superior mobility and surprise are key. An enveloping force should deploy in depth and secure its flanks to avoid being outflanked in turn. Supporting efforts, designed to fix the enemy's attention to his front and forcing him to fight in two or more directions simultaneously, contribute to the main effort's ability to maneuver to the enemy's rear.

Turning Movement

A turning movement is a form of maneuver in which the main effort seizes objectives so deep that the enemy is forced to abandon his position or divert major forces to meet the threat. The intent is to force the enemy out of his position without assaulting him; the act of seizing a key objective to his rear makes his position untenable.

Figure 5-8. Turning Movement.

The main effort usually operates at such a distance from supporting efforts that its units are beyond mutual supporting distance. Therefore, the main effort must be self-sufficient and reach the objective before becoming decisively engaged. Seldom would a turning movement be executed by less than a division.

Distribution of Forces

Sound tactics in the offense are characterized by a concentration of effort against an enemy critical vulnerability, where success will ensure the accomplishment of the mission. The primary way the commander influences the conduct of the attack is through the appropriate distribution of forces into a main effort, one or more supporting efforts, and a reserve.

Main Effort

The commander provides the bulk of his combat power to the main effort to maintain momentum and ensure accomplishment of the mission. The main effort is provided with the greatest mobility and the preponderance of combat support and combat service support. The commander normally gives the main effort priority of fire support. Reserves are echeloned in depth to support exploitation of the main effort's success. The commander can further concentrate the main effort by assigning it a narrower zone of action.

All other actions are designed to support the main effort. The commander disguises the main effort until it is too late for the enemy to react to it in strength. He accomplishes this through the use of demonstrations or feints, security, cover and concealment, and by dispersing his forces until the last instant and achieving mass at the critical time and place. When the main effort fails to accomplish assigned tasks or a supporting effort achieves unexpected success, the commander's C^2 system must facilitate a rapid shift of the main effort.

Supporting Effort

There may be more than one supporting effort. The commander assigns the minimum combat power necessary to accomplish the purpose of each supporting effort. A supporting effort in the offense is carried out in conjunction with the main effort to achieve one or more of the following:

- Deceive the enemy as to the location of the main effort.
- Destroy or fix enemy forces which could shift to oppose the main effort.
- Control terrain that if occupied by the enemy will hinder the main effort.
- Force the enemy to commit reserves prematurely.

Reserve

The reserve is held under the control of the commander as a maneuver force to decisively influence the action. The primary purpose of the reserve is to attack at the critical time and place to ensure the victory or exploit success. Its strength and location will vary with its contemplated mission, the form of maneuver, the terrain, the possible enemy reaction, and the clarity of the situation. When the situation is obscure, the reserve may consist initially of the bulk of the force, centrally located and prepared to be employed at any point. When the situation is clear and enemy capabilities are limited, the reserve may consist of a smaller portion of the force disposed to support the scheme of maneuver. However, the reserve must always be sufficient to exploit success effectively.

The reserve provides the commander the flexibility to react to unforeseen developments. When the reserve is committed, the next higher commander is notified. The reserve should be—

- Positioned to readily reinforce the main effort.
- Employed to exploit success, not to reinforce failure.
- Committed as a maneuver force, not piecemeal.
- Reconstituted immediately.

Conduct of the Offense

The attacker reconnoiters extensively to locate enemy strengths and weaknesses. Once a weakness is identified, the commander rapidly maneuvers his main effort to exploit it. The attacker must minimize
his exposure to enemy fire by using rapid maneuver and counter-fire, exploiting cover offered by the terrain, avoiding obstacles, and maintaining security. The commander makes every effort to achieve surprise by such methods as attacking under cover of darkness or using terrain and/or weather to conceal his force as it closes with the enemy.

The commander directs the battle from a position well forward to develop a firsthand impression of the course of the battle. He personally reallocates resources or shifts his main effort as needed. He provides personal supervision and inspires confidence at key points of the battle. An attack rarely develops exactly as planned. As long as the enemy has any freedom of action, unexpected difficulties will occur. As the attack progresses, control must become increasingly decentralized to subordinate commanders to permit them to meet the rapidly shifting situation.

"One look is worth one hundred reports."
—General Patton citing an
old Japanese proverb

The attacker employs his organic fires and supporting arms to enable him to close with the enemy. The commander prepares for the assault by successively delivering fires on enemy fire suppport assets, C^2 assets and support facilities, and frontline units. These fires protect the attacker and restrict the enemy's ability to counter the attack. Artillery and other supporting arms ensure continuity of support and the ability to mass fires by timely displacement.

During the final stages of the assault, the attacker must rely primarily on organic fires to overcome remaining enemy resistance. The attack culminates in a powerful and violent assault. The assaulting units overrun the enemy using fire and movement. The attacker exploits success immediately by continuing to attack into the depth of the enemy to further disrupt his defense. As the defense begins to disintegrate, the attacker pursues the enemy to defeat him completely.

Decisive victory rarely is the result of success gained in an initial attack; rather, it is the result of quickly and relentlessly exploiting that initial success. The commander seeks to exploit success by constantly pressuring the enemy. As specific opportunities for exploitation cannot be anticipated with certainty, the commander develops sequels based on potential outcomes of the battle. He prepares mentally for any contingency, identifying tentative objectives, zones, concepts of operation, and exploitation forces.

Chapter 6

Defensive Operations

The purpose of the defense is to force the attacker to reach his culminating point without achieving his objectives, to gain the initiative for friendly forces, and to create the opportunity to shift to the offense. The essence of defensive tactics is to place the enemy into a position that permits his destruction through the intelligent use of terrain and firepower, thereby creating a favorable situation for counterattack.

Compared to the offense, the defense is generally the less decisive form of war. While the defense can deny success to the enemy, rarely can it assure victory. In some cases, however, terrain that is critical to the enemy or cannot be bypassed offers the commander an advantage—such advantage that a commander may prefer the defense in order to force the enemy to attack from a disadvantage.

An effective defense is never passive. The defender cannot prepare his positions and simply wait for the enemy to attack. Commanders at every level must seek every opportunity to wrest the initiative from the attacker and shift to the offense. Subordinate commanders take the necessary steps to maintain their positions and cover gaps in their dispositions by the use of observation, obstacles, fires, or reserves. The defense demands resolute will on the part of all commanders.

Defensive Fundamentals

The GCE commander considers the following fundamentals when conducting defensive operations.

Maneuver

Maneuver is as important in the defense as it is in the offense. While steadfastness and the tenacious holding of key terrain is essential in the defense, the defender must not become immobile. The defender must maintain his freedom of maneuver. Maneuver is essential in generating the offensive power fundamental to a successful defense. Maneuver is essential to security operations, operations within the main battle area, and rear operations. Units of all sizes maneuver in depth, taking advantage of terrain and tactical developments, to concentrate, disperse, and occupy positions from which they can bring more effective fire to bear on the enemy.

Preparation

The defender usually organizes the defense on terrain of his choosing. While the attacker can choose the specific time and point of attack, the defender, through the proper selection of terrain and reinforcing obstacles, can direct the energy of the enemy's attack into terrain which is advantageous to the defender. The defender must take advantage of this by making the most thorough preparations that time allows. Preparations should begin as early as possible and continue throughout the battle. It must be understood that these preparations may be under constant observation by the attacker. To inhibit the enemy's intelligence effort, the commander establishes security forces to conduct counter-reconnaissance and deceives the enemy as to the exact location of the main defenses.

The commander's organization of the ground consists of his plans for fires and maneuver; counterattack plans; and preparation of positions, routes, obstacles, logistics, and C^2 facilities.

Mass and Concentration

The defender cannot defend everywhere in strength. He must concentrate forces and fires at the decisive place if he is to succeed, while exercising economy of force in less critical areas. Some portions of the front may be unoccupied but held effectively by a combination of fire and obstacles. Additionally, security forces, sensors, and surveillance can be employed along less likely avenues of approach to help reduce risks.

The commander designates his main effort based on his anticipation of the enemy's main effort. The defensive scheme of maneuver takes advantage of terrain essential to the integrity of the defense. Reserves are positioned so that they can intervene quickly to support the main effort.

Since he usually cannot determine with certainty where the enemy will attack, the commander must be prepared to quickly shift his main effort. The defender masses fires and concentrates combat power repeatedly to wrest the initiative from the attacker. The commander does this swiftly, since periods that allow him to develop superior combat power will be brief. The commander may have to surrender some ground to gain the time necessary to concentrate forces.

Flexibility

"Petty geniuses attempt to hold everything; wise men hold fast to the key points. They parry great blows and scorn little accidents. There is an ancient apothegm: he who would preserve everything preserves nothing."

—Frederick the Great

While the commander makes every effort to determine the enemy's intentions in advance, the plan must be flexible enough to deal with different enemy courses of action. Flexibility is created by—

- Detailed planning for contingencies.
- Designating supplementary and alternate positions.
- Properly locating, task organizing, and planning use of the reserve.
- Designing counterattack plans.
- Preparing to assume the offense.
- Planning on-call fire support.

Offensive Action

Since the offense is the decisive form of combat, the commander seeks every opportunity to take offensive action. The defender takes offensive action by—

- Launching spoiling attacks while the enemy is preparing or assembling for an attack.
- Attacking with security forces to harass, distract, deceive, and damage the enemy before he reaches the main battle area.
- Counterattacking to destroy or repulse enemy penetrations.

Counterattacks range from immediately executed actions for reestablishing the integrity of the defense to commitment of the reserve at the decisive moment of the battle. The commander prepares to change to the offense at the earliest feasible opportunity.

"In war, the only sure defense is offense . . . ."
—General Patton Use of
Terrain

The defender must exploit every aspect of terrain and weather to his advantage. In the defense, as in the attack, terrain is valuable only if a force gains advantage from its possession or control. In making his estimate of the situation, the defending commander takes account of key terrain and visualizes all possible enemy avenues of approach into the sector. The defender seeks to defend on terrain that maximizes effective fire, cover, concealment, movement, and surprise. A position combining all these defensive advantages will seldom be available. While capitalizing on the strong points of the terrain, the defender strengthens the weak points. Natural obstacles are exploited and reinforced by the defender.

Obstacle integration multiplies the effects and capabilities of firepower. The combination of firepower and obstacles causes the enemy to conform to our scheme of maneuver. Obstacles magnify the effects of firepower by increasing target acquisition time and by creating exploitable vulnerabilities. Obstacles not properly integrated with maneuver and the plan of supporting fires inhibit friendly maneuver and waste resources and have a negligible effect on enemy maneuver.

Mutual Support

Mutual support strengthens any defense. Mutual support is that support which units render each other. Mutual support is achieved when defensive positions are located in such a way that the enemy cannot attack one position without coming under fire from at least one other. The degree of mutual support obtained depends on the terrain, range of weapons, and visibility. Ideally, the frontage a force must defend is directly related to its ability to provide mutual support between its units. To neutralize mutually supporting positions, an attacker must disperse fire away from his main objective. Mutual support is essential at all levels.

Defense in Depth

Defense in depth is the siting of mutually supporting defensive positions throughout the main battle area to absorb and progressively weaken the attack. It provides maneuver space within the defensive sector for the maneuver of subordinate units against the enemy's main effort. Defense in depth is necessary to—

- Disrupt the momentum of the attack and prevent a breakthrough.
- Force the enemy into engagement areas.
- Allow the defender time to determine the enemy's main effort and to counter it.
- Force the enemy to commit his reserves at a nondecisive point.
- Disperse the effects of enemy fire.

The greater the enemy's combat power and the wider the frontage held, the greater the depth of the defense must be. Defense in depth is achieved by—

- Engaging the enemy at the earliest opportunity with security forces.
- Employing weapons at maximum effective range.
- Using blocking positions, obstacles, and supplementary positions throughout the main battle area.
- Positioning and moving reserves and fire support units.

Surprise

The defense, no less than the offense, must achieve surprise. The organization of a defense must not betray the commander's intent and positioning of units. To preserve combat power, especially against a superior enemy, the defender must employ every means to mislead the enemy as to the true location of his positions and as to the strength and disposition of forces. Toward this end, the commander considers the use of security forces, C^2W, and reverse slopes and maximizes available cover, concealment, camouflage, and dummy positions. The best defensive terrain will likely also be apparent to the attacking enemy, who will maneuver against it with caution and will mass fires on it. When possible, the commander selects terrain that has good defensive qualities but is not conspicuous.

Knowledge of the Enemy

The defense is largely reactive by nature. A defender's options are dictated in large part by what the attacker does. Therefore, thorough knowledge of the enemy's capabilities, operational concepts, and habits is essential to a successful defense. The defending commander must look at his force and his sector through the enemy's eyes to identify probable enemy objectives and courses of action. A thorough IPB will provide valuable indications of enemy assembly areas, attack positions, routes, firing positions for supporting arms units, axes of advance, and the area most advantageous for the main effort. When the defender can accurately anticipate the enemy's actions, he can trap the attacker within the defense and establish conditions for resumption of offensive operations.

Organization of the Defense

The defensive sector is organized in depth into three areas: the security area, main battle area, and rear area. See figure 6-1.

Figure 6-1. Organization of the Defense.

Security Area

For any echelon of command, the security area is the area forward of the forward edge of the battle area (FEBA) assigned to the security forces. It is here that security forces execute assigned tasks. The commander adds depth to the defense by extending the security area as far forward as is tactically feasible. This allows security forces to inflict the greatest possible damage and disruption to the enemy attack by the time the enemy reaches the main battle area. Normally, the commander extends the lateral boundaries of subordinate units forward of the FEBA, giving those units responsibility for the security area within sector to the forward extension of their boundaries.

Main Battle Area

The main battle area is the area extending from the FEBA to the rear boundaries of the forward subordinate units. The commander positions forces throughout the main battle area to destroy or contain enemy assaults. Reserves are employed in the main battle area to reduce penetrations, regain terrain, or destroy enemy forces. The greater the depth of the main battle area, the greater the maneuver space for fighting the main defensive battle afforded subordinate commanders. It is in the main battle area that the decisive defensive battle is usually fought.

Rear Area

The rear area is the area extending forward from a command's rear boundary to the rear boundary of the main battle area. This area is provided primarily for combat service support functions. Rear operations include those functions of security and sustainment required to maintain continuity of operations by the force as a whole.

Distribution of Forces

The defender organizes his force as follows: security forces, main battle forces, and reserves.

Security Forces

Security forces are employed forward of the main battle area to delay, disrupt, and provide early warning of the enemy's advance and to deceive him as to the true location of the main battle area. These forces are assigned cover, guard, or screen missions. Operations of the security forces must be an integral part of the overall defensive plan. When subordinate units are assigned a portion of the security area in the conduct of the higher organization's security operations, they establish the appropriate security force under a single commander within the assigned zone of action or sector. Commanders of these subordinate unit security forces respond directly to taskings by the commander assigned overall responsibility for the higher organization's security operation. The mission assigned those forces is based on the situation. At each higher echelon, available resources allow the security force to operate at a greater distance forward of the main battle area.

Covering Force

The GCE may provide the bulk of the MAGTF's covering force. The covering force operates apart from the main force to engage, delay, disrupt, and deceive the enemy before he can attack the main force. It accomplishes this by conducting offensive and/or defensive operations. The size of the covering force is METT-T-dependent and may contain tanks, light-armored vehicles, artillery, assault amphibian vehicles with embarked infantry, engineer, and combat service support assets. The covering force may be controlled by the GCE, the MAGTF, and in some instances, the ACE, depending on the situation.

Guard Force

The GCE may designate a guard force for protection from enemy ground observation, direct fire, and surprise attack for a given period of time. A guard force allows the commander to extend the defense in time and space to prevent interruption of the organization of the main battle area. Observation of the enemy and reporting of information by the guard force is an inherent task of the guard force, but secondary to its primary function of protection. A guard force is organized based upon METT-T. The GCE commander determines the orientation of the guard force and the duration the guard must be provided. Normally, guard forces are oriented to the flanks for the minimum amount of time necessary to develop an integrated defense. When the GCE commander determines that the requirement for a guard force has expired, the guard force may receive a cover or screen mission with the requisite loss or gain of resources.

Screening Force

The GCE may establish a screening force to gain and maintain contact with the enemy, to observe enemy activity, to identify the enemy main effort, and to report information. In most situations, the minimum security force organized by the GCE is a screening force. Normally, the screening force only fights in self-defense, but may be tasked to—

- Repel enemy reconnaissance units as part of the GCE's counter-reconnaissance effort.
- Prevent enemy artillery from acquiring terrain that enables frontline units to be engaged.
- Provide early warning.
- Attack the enemy with supporting arms.

Local Security

All units of the GCE provide local security. The depth of local security is dictated by terrain, communications, target acquisition capabilities, and the enemy threat.

Passive Security Measures

All units employ passive security measures to reduce exposure to the enemy, to include observation, electronic exposure, and fires. Communications procedures, camouflage, movement control, and other individual continuing actions are strictly enforced.

Active Security Measures

Active security measures are employed by the GCE and coordinated at all levels. Active security measures include combat patrolling, sensors, target acquisition radars, surveillance, and employment of false visual and electronic signatures. In addition, skills of certain units within the GCE enhance the security posture of the organization. For example, engineers within the GCE contribute to survivability, mobility, and countermobility, all of which contribute to security. Any active measure that may impact on other elements of the MAGTF is coordinated throughout the MAGTF.

Main Battle Forces

Main battle forces engage the enemy in decisive combat to slow, stop, canalize, disorganize, and defeat his attack. Main battle forces occupy defensive positions within the main battle area. Positions are oriented on the most likely and most dangerous avenues of approach into the sector. Forces responsible for the most dangerous approach are normally assigned the initial main effort. The commander can strengthen his defense at this point by narrowing the sector of and providing the priority of support to the unit astride it.

Reserves

The reserve is a part of the force, held under control of the commander as a maneuver force to influence the action. Missions assigned to the reserve normally consist of counterattack, reinforcement of the main effort, protection of flanks, and supporting committed units by fire. Until employed, reserves normally occupy covered assembly areas in the rear area, protected from enemy fires.

Types of Defensive Operations

Every defense contains two complementary characteristics: a static —or positional—element, which anchors the defense to key terrain; and a dynamic—or mobile—element, which generates combat power through maneuver and concentration of forces. The positional element is characterized by use of battle positions, strongpoints, fortifications, and barriers to halt the enemy advance. The mobile element is characterized by the use of offensive action, supplementary positions, planned delaying actions, lateral shifting of forces, and commitment of the reserve. Conceptually, this results in two defensive extremes: the position defense and the mobile defense. However, neither type can be used exclusively in practice; although these descriptions convey the general pattern of each type of defense, any defense will include both positional and mobile elements.

Commanders may conduct position and mobile defenses simultaneously to take advantage of the strengths of subordinate organizations. Units with significant mobility may be designated part of the reserve or tasked to conduct mobile-type defenses, given the situation and terrain within their assigned sector. Other units without a mobility advantage over the enemy force and given the nature of the terrain may be assigned a position defense mission. Irrespective of the type of defense employed, the defender must conduct a decisive counterattack or resume the offensive once the enemy is defeated or reaches his culminating point.

Position Defense

The position defense is conducted to deny the enemy access to critical terrain for a specified period of time. The bulk of the defending force is deployed in a combination of static defense and small, mobile reserves. Mutual support and positions in depth force the enemy to expose his force in the attack of each position.

Principal reliance is placed on the ability of the forces to maintain their positions and to control unoccupied terrain by fire. The reserve is used to blunt and contain penetrations, to reinforce the main effort, and to counterattack to destroy enemy forces.

The position defense is seldom capable of achieving the outright destruction of the attacking force due to its limited mobility. The attacker may disengage when dealt a tactical setback or take advantage of other opportunities to maintain the initiative. Thus, the position defense relies on other simultaneous or subsequent operations by adjacent or reinforcing forces to achieve decisive results. Circumstances may require or favor the conduct of a position defense when—

- Specific terrain is so militarily or politically critical it must be defended.
- The defender possesses less mobility than the enemy.
- Maneuver space is limited.
- Terrain restricts the movement of the defender.
- Terrain permits surprise fires to be massed on the bulk of the enemy force.
- Terrain does not permit the attacker mutual support.

Mobile Defense

The mobile defense orients on the destruction of the enemy through offensive action. The bulk of the force is held as a mobile striking force with strict economy applied to dedicated positional supporting efforts designed to canalize, delay, and disrupt the attack. The commander can then take advantage of vulnerabilities created in the enemy's effort to defeat the positional elements of the defense.

C^2W and our interpretation of the enemy's operational intent are used to focus the enemy on a noncritical objective and then to counterattack him from an unexpected direction. Mobile defense also requires effective counter-reconnaissance coupled with recognition of enemy C^2 nodes, sustainment elements, and fire support units.

This combination of assets and information allows the GCE commander to blind the enemy, then strike throughout the depth of the enemy force at the decisive time and place.

To succeed, the mobile element of the defense must have mobility greater than that of the enemy. Terrain is traded in order to extend the enemy and expose his flanks and allow the defender to maximize the benefit of the terrain for purposes of counterattack. To draw the enemy into an engagement area, a mobile defense requires depth.

Forms of Defensive Maneuver

There are two broad forms of defensive maneuver—defend and retrograde. Most defensive schemes of maneuver will incorporate a combination of these two forms. A unit that defends accomplishes this through the assignment of battle positions, blocking positions, sectors, and strongpoints. Retrograde includes delaying the enemy, withdrawal, and retirement.

Defend

Subordinate units that defend do so through the assignment of sectors, battle or blocking positions, and strongpoints. These assignments are made in a manner that enhances depth and mutual support; that provides opportunities to trap or ambush the attacker; and that affords observation, surprise, and deception. Defending units must also maintain an awareness of concurrent delaying actions to take advantage of opportunities created by adjacent units. Defend missions are not assigned solely to main battle area units and may be required of units in the security and rear areas.

Sector

Assignment of defensive sectors to subordinate commanders provides these commanders with maximum latitude to accomplish assigned tasks. The extent of the sector assigned is METT-T dependent, but as a general rule should be no larger than can be influenced by the unit. Within his assigned sector, the commander may assign subordinates sectors, battle positions, strongpoints, or any of these in combination.

Battle Position

A battle position is a defensive position from which a unit will fight. The unit may vary in size from a platoon to a battalion. Local security may operate outside the battle position for early detection of the enemy and all-around security. Battle positions may be occupied hastily and may be held only temporarily but should be improved continuously while occupied. A blocking position is a battle position so sited as to deny the enemy access to a given area or to prevent his advance in a given direction.

Strongpoint

A strongpoint is a strongly fortified defensive position designed to defeat enemy armor and mechanized attacks. A strongpoint is located on a terrain feature that is critical to the overall defense and is intended to be occupied permanently or for an extended period of time. A strongpoint normally is occupied by a company or larger organized for all-around defense. A unit or organization holding a strongpoint may be cut off and lose its freedom of maneuver, so it should have its own CSS. A strongpoint is established only after the commander determines that a position must be retained at all costs.

Retrograde

A retrograde operation is a movement to the rear or away from the enemy. A retrograde may be a planned movement or one forced by enemy action. Retrogrades may be classified as delay, withdrawal, or retirement.

Delay

A delay is an operation in which a force under pressure trades space for time by slowing down the enemy's momentum and inflicting maximum damage on the enemy without becoming decisively engaged. The commander of the overall defense must specify the amount of time to be gained by the delaying force to successfully accomplish the mission. Time may be expressed in hours or in events to be accomplished by the main battle area forces. Delays may be used appropriately in the security area, main battle area, or rear area. Sufficient depth of area is required for a delay. Delays are conducted—

- When the force's strength is insufficient to defend or attack.
- To reduce the enemy's offensive capability by inflicting casualties.
- To gain time by forcing the enemy to deploy.
- To determine the strength and location of the enemy's main effort.
- When the enemy intent is not clear and the commander desires intelligence.
- To protect and provide early warning for the main battle area forces.
- To allow time to reestablish the defense.

Fundamentals of the Delay

In the delay, decisive engagement is to be avoided. Special consideration is given to the following fundamentals:

Centralized control and decentralized execution: A delaying action is normally characterized by operations on a wide front with most of the delaying forces forward and minimum forces in reserve. This results in a series of independent actions across the sector in which each commander must have freedom of action while engaging the enemy.

Maximum use of terrain and obstacles: Obstacles are emplaced and natural obstacles are enhanced to canalize and delay the enemy. Blocking positions may be incorporated in the delay and located on terrain dominating avenues of approach that provide long-range fields of fire and facilitate disengagement.

Maximum use of fires: Long-range fires, to include offensive air support, are brought to bear against the enemy to destroy high-payoff targets and to force the enemy to deploy. Long-range fires must be thoroughly planned and coordinated by the GCE and MAGTF to ensure unity of effort and proper utilization of all available Marine and joint fire support means.

Force the enemy to deploy and maneuver: Delay forces must be strong enough to force the enemy to deploy prematurely, thus slowing his advance. Strong delay forces can also disguise the location of the main battle area, disguise the GCE's main effort, and help identify the enemy's main effort.

Maintain contact: Maintaining contact with the enemy prevents him from bypassing delaying forces; advancing unimpeded by forcing him to maintain his combat formations; slows his mobility by forcing him off high speed avenues of approach; and provides continuous information to the GCE commander.

Avoid decisive engagement: Units decisively engaged lose their freedom of maneuver and must fight the engagement to a decision. Consequently, they lose the ability to continue to accomplish the assigned delay mission.

Techniques for Delaying

Units conducting a delay can delay from successive or alternate positions, or a combination of the two. The method adopted depends largely on the width of the assigned sector and the forces available.

Delay from successive positions: This is a technique for delay in which all delaying units are positioned forward in a single echelon. This technique is appropriate for delaying in less threatened areas or against enemy supporting efforts. It is normally used when the terrain within the sector is favorable to enemy maneuver and the delaying force has greater mobility than the attacker. Units delay continuously on and between battle positions throughout their sectors, fighting rearward from one position to the next, holding each position for a specified period of time or as long as possible. See figure 6-2.

Figure 6-2. Delay from Successive Positions.

Delay from alternate positions: This is a technique of delay in which a unit delays in sector with subordinate units deployed in depth. As the forward unit delays, another subordinate unit occupies the next position in depth and prepares to assume the fight. The forward unit disengages and passes through the next rearward position and prepares for subsequent delay at the third position in depth after handing over the battle to the second unit. This technique may be used when the sector is narrow enough to permit the deployment of the force in depth; when terrain restricts enemy maneuver; or when the delaying force has less mobility than the attacker. See figure 6-3.

Withdrawal

A withdrawal is a planned operation in which a force in contact disengages from an enemy force. The commander's intention in a withdrawal is to put distance between his force and the enemy as quickly as possible and without the enemy's knowledge. A withdrawal may be undertaken—

- If the objective of the operation cannot be achieved and the force is in danger of being destroyed.
- To avoid battle under unfavorable conditions.
- To draw the enemy into an engagement area.
- To allow for the use of the force or parts of it elsewhere.

There are two types of withdrawal—a withdrawal under pressure and a withdrawal not under pressure. A prudent commander always attempts to conduct a withdrawal not under pressure, but plans to execute the withdrawal under pressure. See figures 6-4 and 6-5.

Figure 6-3. Delay from Alternate Positions.

Figure 6-4. Withdrawal Under Pressure

Figure 6-5. Withdrawal Not Under Pressure

Retirement

A retirement is an operation in which a force out of contact moves away from the enemy. A retiring unit normally is protected by another unit between it and the enemy. However, the retiring unit must establish security. Often a retirement immediately follows a withdrawal.

Planning for the Defense

Planning for the defense begins when the commander receives a mission or warning order to defend or anticipates a need to do so. To facilitate preparations, concurrent planning at all levels of command is essential. The defensive plan should accentuate the natural strengths of the terrain and the defending force. The defensive plan consists of a scheme of maneuver and a plan for supporting fires.

Scheme of Maneuver

The scheme of maneuver for the defense includes—

- Initial positions, withdrawal routes, and passage points for the security force.
- Primary, alternate, and supplementary positions for main battle area forces.
- Counterattack plans.
- Contingency plans to block penetrations or reinforce threatened areas.
- Dummy positions designed to deceive the enemy.
- Planned retrogrades to draw the enemy into engagement areas.
- Obstacles and barriers integrated with the scheme of maneuver and fire support plan.

Fire Support Plan

The fire support plan must support the scheme of maneuver. It is normally designed to place the enemy under increasing volumes of fire as he approaches a defensive position. Deep supporting fires are delivered by aircraft and long-range indirect fire weapons. Fires are planned along expected enemy routes, in engagement areas, around obstacles and barriers, and within the defensive positions. The degree of completeness and centralization of defensive fire planning depends on the time available to prepare for the defense. Ordinarily, defensive fire support plans are based on terrain, friendly positions, and barriers. Close supporting fires are closely integrated with infantry, tank, and antitank direct-fire weapons.

Preparing for the Defense

Any time the commander is not engaged in the attack, he must initiate preparations for the defense. Similar to the attack, the type of defense depends on preparation time and is considered either deliberate or hasty. Preparations are made simultaneously at all levels of command. The more carefully and comprehensively the defense is prepared, the stronger it will be. Priorities of work must be established.

Deliberate Defense

A deliberate defense is a defense normally organized when out of contact with the enemy or when contact with the enemy is not imminent and time for organization is available. A deliberate defense normally includes fortifications, strongpoints, extensive use of obstacles, and fully integrated fires. The commander normally is free to make a detailed reconnaissance of his sector, select the terrain on which to defend, and decide the best tactical deployment of forces.

Hasty Defense

A hasty defense is a defense normally organized while in contact
with the enemy or when contact is imminent and time available for
organization is limited. Reconnaissance of the sector must be cur-
tailed and the defense assumed directly from the current positions
of units. Depending on the situation, it may be necessary for a
commander to initiate a hasty attack to seize terrain suitable to his
defense. Or, the commander may employ a security force to delay
the enemy while deploying the bulk of his force to more suitable
defensive terrain. A hasty defense is improved continuously as the
situation permits and may eventually become a deliberate defense.

Conducting the Defense

Security forces at all levels warn of the enemy approach. Within
their capabilities, the security forces proceed to strip away enemy
reconnaissance and security elements. These forces then deceive
the enemy as to the true location of the main battle area and principal defensive positions. Finally they disrupt, delay, and damage
the enemy as much as possible without becoming decisively engaged. The ultimate goal of security forces is to destroy as much
of the enemy as possible, within their capabilities, without losing
the freedom to maneuver, to prevent surprise, and to provide time
to main battle area forces.

At a predetermined location, control of the battle is transferred to
security elements established by the next subordinate command.
This location is known as a handover line. A handover line is a
control feature, preferably following easily defined terrain features,
used to pass responsibility for the conduct of combat operations
from one force to another. This transfer of control must be
carefully coordinated. The security forces conduct a rearward
passage

of lines at designated passage points, and the senior command's security force withdraws in preparation for its subsequent mission. The subordinate unit's security force supports the disengagement of the senior command's security force.

Security forces in one part of the security area do not withdraw automatically because adjacent forces have been forced rearward. While adjusting to the enemy advance and securing its flanks to avoid being cut off, security forces should continue their mission when possible. Retaining forward positions in part of the sector provides surveillance and control of supporting arms into the enemy's depth, allows the commander to concentrate temporarily on a narrower front in the main battle area, and provides access to the enemy's flank for a counterattack.

The defender engages the enemy with long-range fires as early as possible unless fires are withheld to prevent the loss of surprise. Commanders make maximum use of fire support to destroy and disrupt enemy formations as they approach the main battle area. As the enemy closes, he is subjected to an ever-increasing volume of fires from the main battle area forces and all supporting arms. Obstacles and barriers are located to delay or canalize the enemy and are covered by fire to destroy him while he is halted or slowed and concentrated on the process of breaching. Main battle forces maintain an offensive spirit throughout the battle, executing local counterattacks whenever there is a probability of success.

Combat power that can be concentrated most quickly, such as offensive air support and artillery, is brought to bear while tanks and infantry move into position. The defender reacts to the enemy's main effort by reinforcing the threatened sector or allowing the enemy's main effort to penetrate into engagement areas within the main battle area, then, cutting him off and destroying him by counterattack. When the enemy attack has been broken, the commander looks to exploit any advantageous situations.

Use of the Reserve in the Defense

The weaker the defender, the more important the defender's reserve becomes. The less that is known of the enemy or his intention, the greater the proportion of combat power that must be held in reserve. The commander withholds his reserve for decisive action and refuses to dissipate it on local emergencies. Once a reserve has been committed, a new reserve must be created or obtained. The reserve provides the defender flexibility and balance to strike quickly with concentrated combat power on ground of the defender's choosing.

Reserves must be organized based on METT-T. The tactical mobility of mechanized and helicopterborne forces makes them well suited for use as the reserve. Mechanized reserve forces are best employed offensively. In suitable terrain, a helicopterborne reserve can react quickly to reinforce main battle area positions or block penetrations. However, helicopterborne forces often lack the shock effect desired for counterattacks.

Timing is critical to the employment of the reserve. As the area of probable employment of the reserve becomes apparent, the commander moves his reserve to have it more readily available for action. The commander's intent must clearly state the circumstances calling for the commitment of the reserve. When he commits his reserve, the commander must make his decision promptly and with an accurate understanding of movement and deployment times. If committed too soon or too late, the reserve may not have a decisive effect.

To conserve forces, the commander may choose to use security forces as part or all of his reserve after completion of their security mission. However, the commander must weigh the decision to do this against the possibility that the security force may suffer a loss of combat power during its security mission. The loss of combat power may reduce the capability of the security force to accomplish its subsequent mission as the reserve.

Chapter 7

Operations Other Than War

Operations other than war (OOTW) encompass a wide range of activities where the military instrument of national power is used for purposes other than the large-scale combat operations usually associated with war. Although these operations are often conducted outside the U.S., they also include military support to U.S. civil authorities. They may be regional in nature, may develop quickly, and may or may not be long term. A MAGTF GCE employed in response to a crisis serves to contain or limit its immediate effects and strives to achieve the peaceful resolution of the issues that created it. There are two broad categories of OOTW based on the general goal—*operations that deter war and resolve conflict* and *operations that promote peace*. See figure 7-1.

The forward-deployed MAGTF integrated with the NEF is uniquely capable of conducting OOTW. Though trained and equipped primarily for combat, the MAGTF can be task organized to meet the mission requirements of the contingency at hand—from direct combat against a capable enemy force to the civil assistance necessary to maintain the basic essentials of life for a disaster-stricken populace.

OOTW are often conducted in a politically sensitive environment. Marines must consider every individual action as it may have significant political or operational impact. This places increased importance on small-unit discipline, decentralized execution of lawful orders, cultural training, and proper use of individual language

Figure 7-1. Range of Military Operations.

capabilities within the force. For example, one act of civil distur-
bance or intolerant treatment of civilians can turn a supportive
populace against the force and be exploited by a potential adversary. This same act may become a lighting rod in turning domestic
public opinion against a continued effort.

The commander must consider his activities in relation to similar
activities carried out by agencies of the U.S. government, allies,
and the host nation, as well as nongovernment and private volunteer
organizations. Additional considerations include the following:

- Media scrutiny will be extensive.
- Rules of engagement will be more restrictive.
- Identification of hostile parties may be more difficult.

- Military assets may be routinely used to support noncombat functions.
- Interaction with civilian noncombatants will be routine at every level of command.

Mission-essential training for combat will also prepare organizations for these operations as will small-unit leadership and discipline. A fully combat-trained Marine can function in any OOTW. However, the environment and circumstances in which these operations are conducted will vary significantly from that normally associated with combat operations. Training specifically designed for the conduct of unique OOTW missions may be a luxury due to the expected lack of time and scarcity of resources. Furthermore, the commander must understand that a protracted OOTW may cause a degradation in the GCE's capability to conduct its primary mission of combat due to the lack of training opportunities during the operation. GCE training for combat operations relating to OOTW include the following:

- Raids (all types).
- Noncombatant evacuation operations (NEO) (permissive and nonpermissive).
- Military operations on urbanized terrain (MOUT). (Urban areas are normally the focus of political, ethnic, economic, and religious conflict and strife.)
- Security operations.

Principles of Operations Other Than War

Commanders should consider the following principles when they are planning and conducting OOTW. The principles of OOTW must be carefully applied when analyzing the requirements of the mission because of the nature of OOTW and the impact an error in judgement can have on an operation. Commanders and all Marines should have a broad and intricate understanding of these principles to ensure the force is properly prepared for the demands of OOTW.

Objective

A clearly defined and attainable objective, with a precise definition of success, is critical. Multinational forces must come to a unanimous agreement as to what the objective is, recognizing that individual nations may want to achieve it by vastly different means. Leaders of different organizations, military and otherwise, must understand the strategic aims, set appropriate objectives, and ensure that they contribute to overall unity of effort.

Unity of Effort

Unity of effort is similar to unity of command associated with combat operations. In OOTW, unity of effort may be more difficult to attain because ad hoc alliances, coalitions, and the introduction of nonmilitary agencies will pose unique problems of coordination and cooperation. Organizations such as the United Nations, NATO, the State Department, United States Agency for International Development (USAID), and other regional alliances establish the political, economic, military, and psychological atmosphere of the operation. The MAGTF and GCE will normally support the efforts of these organizations and the host nation. Multinational command relationships may be loosely defined. This will require senior military and political decisionmakers to be on the scene as early as possible and commanders to emphasize cooperation and coordination rather than command authority to achieve objectives.

In multinational coalitions, even if unity of command is established, unity of effort may still be elusive. The GCE commander must understand that participating forces will be compelled to take direction from their own national authorities and respond to their own national interests. Likewise, most participating forces receive logistic support through their own sustainment system.

Security

Security deals primarily with force protection and protection of civilian noncombatants. The presence of military forces may generate opposition by various elements that adhere to different social, political, or military objectives. These factions might attack the force to gain legitimacy, to weaken U.S. or international resolve, or to undermine the authority of the host government. U.S. forces are a particularly desirable target as they represent the world's sole superpower. U.S. forces may have difficulty appearing as impartial under a scrutinizing media. Protected parties may be perceived as supporting or supported by the U.S. government. This perception could place the protected party at greater risk. Marine forces must maintain constant vigilance regardless of their apparent acceptance by the populace. They must be ready to exercise their inherent right to self-defense by rapid transition from peaceful activities to a combat posture.

Restraint

Rules of engagement (ROE) are established by the commander in chief (CINC) and based on National Command Authorities (NCA) guidance, mission, threat, law of war, and host nation restraints on force deployment. These rules are communicated to the GCE through the chain of command. ROE must be thoroughly briefed, understood, and enforced by all Marines. The use of force is characterized by the judicious and prudent selection, deployment, and employment of forces most suitable to the situation. This never prevents units from exercising their inherent right to self-defense or the application of appropriate combat power to demonstrate U.S. resolve.

Changes to the ROE are made by requesting supplemental guidance through the chain of command. Meanwhile, the local commander should publish his own unclassified "Commander's Guidance on the Use of Force" to ensure that every individual understands the restrictions. Intelligence also plays an important part in developing ROE as required by the changing situation. Anticipation of unforeseen conditions and timeliness of getting changes approved is critical.

For OOTW, commanders develop their guidance with the following in mind:

- Explain the mission and commander's intent.
- Assess the threat accurately.
- State guidance clearly.
- Recognize that use of force is justified in self-defense.
- Anticipate that guidance is subject to change.

Legitimacy

Legitimacy of the operation and host government must be maintained. During operations where a legally constituted government does not exist, extreme caution must be applied when dealing with indigenous individuals and organizations. The appearance of formal U.S. recognition when such U.S. support does not exist must be avoided. Commanders should incorporate psychological operations/public affairs programs in their planning to influence both foreign and domestic perceptions. Activities that attempt to solve immediate problems yet undermine the authority or legitimacy of the host government may undermine our ultimate aim.

Operations to Deter War and Resolve Conflict

In spite of internal or external efforts to promote peace, factions within a country or region may resort to armed conflict. A deteriorating situation may require military force to demonstrate U.S.

resolve and capability, to support the other instruments of national power, or to terminate the situation on favorable terms. The general goals of U.S. military operations during periods of conflict are to support national objectives, deter war, and return to a state of peace. These operations involve a greater risk as they may escalate into large-scale combat operations. Operations to deter war and resolve conflict include *support to insurgency and counterinsurgency, combating terrorism, show of force, noncombatant evacuation operations, recovery, attacks and raids, maritime interception operations,* and *peace operations.*

Support to Insurgency and Counterinsurgency

The U.S. may support insurgencies that share U.S. values. It may also support counterinsurgencies of friendly governments against insurgents that proclaim support of ideology incompatible with U.S. national interests. Fundamental to supporting insurgencies or counterinsurgencies is the recognition of the political, economic, and/or ideological motivation of the insurgent movement. Leaders must understand the culture of the population and the geographical nature of the country or countries involved. This understanding is critical to the decision to commit U.S. forces, to determining the extent of the military operation, and to identifying the threat the insurgency poses to our national interests. The wide range of capabilities within the MAGTF are valuable in supporting major insurgencies and counterinsurgencies.

Support to Insurgency

Since most U.S. support to insurgencies is covert, MAGTF support may be limited to supporting the efforts of special operations forces. GCE support to insurgencies will principally involve training and advising insurgent forces in unconventional warfare tactics, techniques, and procedures. Insurgency support is classified as a special activity and is subject to approval by the U.S. Congress.

Support to Counterinsurgency

The MAGTF GCE may be tasked to provide support across the full range of conventional capabilities to the supported government against a hostile insurgent force. The GCE may be tasked by the MAGTF CE to support counterinsurgencies by assisting with foreign internal defense; training of military and law enforcement personnel; and the conduct of strikes, raids, and limited ground combat. Though the GCE may participate in combat operations in support of a friendly country's counterinsurgency effort, in many instances, this active role may detract from the political legitimacy and effectiveness of the host government. Therefore, the employment of the GCE and its contribution to the host nation must be continually assessed based on the changing situation. The spectrum of support provided by the GCE includes—

- Advisory and training assistance.
- Intelligence support.
- Logistics support.
- Civil-military operations (CMO).
- C^2 support.
- Combat operations.

Combating Terrorism

Terrorism is designed to influence public support for a stated policy or program by successful attacks against citizens and property. Terrorist attacks may reduce the credibility of the U.S. or diminish the nation's ability to influence international events. The lead agency for combating terrorism overseas is the Department of State and, within the CONUS, the Department of Justice. The Department of Defense is responsible for providing technical assistance and/or forces when directed or requested by one of these lead agencies.

Since terrorism can be international in scope and, in some instances, aided and abetted by state sponsors, the threat posed to U.S. citizens and security interests abroad may require a U.S. military response. The two levels of response are categorized as *counterterrorism* and *antiterrorism.*

Counterterrorism

Counterterrorism is the offensive portion of combating terrorism. It involves the employment of forces to directly address terrorist situations including preemptive, retaliatory, and rescue operations. The extent of forward-deployed MAGTF participation depends upon location, type of incident, the degree of force required, and the impact of legal and political constraints. National assets are normally used to conduct counterterrorism operations. A forward-deployed GCE within striking distance of a terrorist action may be tasked to support an in-extremis effort.

Antiterrorism

For a discussion of antiterrorism, see page 7-15.

Show of Force

Show of force lends credibility to U.S. policies and commitments, increases its regional influence, and demonstrates resolve. Forward-deployed naval expeditionary forces have historically been the instrument of a show of force. Additionally, combined training exercises and ship and aircraft visits can also influence other governments or organizations to respect U.S. interests and international law.

Noncombatant Evacuation Operations

Noncombatant evacuation operations (NEO) remove threatened civilian noncombatants from locations in a foreign nation. The methods and timing for the evacuation will be significantly influenced by diplomatic considerations. NEO may also entail the evacuation of U.S. citizens and/or citizens of nonbelligerent countries and the host nation. Ideally, there is no opposition to an evacuation, and it will be supported by the host country. However, commanders must be prepared to conduct a NEO in a hostile environment that requires the use of force by the MAGTF.

NEO resemble raids in that they involve the swift introduction of forces, evacuation of the noncombatants, and a planned withdrawal. Detailed coordination is required between the MAGTF and the representatives of the Department of State responsible for U.S. interests in the region. The U.S. ambassador, or chief of the diplomatic mission, is responsible for the preparation and update of the regional emergency action plan that covers the military evacuation of U.S. citizens and other designated foreign nationals. Execution of the military portion of the emergency action plan is the sole responsibility of the supporting military commander.

Recovery

Recovery operations are sophisticated activities requiring detailed planning and precise execution, especially when conducted in hostile areas. When conducted by the military, they may be clandestine or overt. These operations may include the recovery of U.S. or friendly foreign nationals or sensitive equipment critical to U.S. national security. Recovery operations may be conducted by specially trained units or conventional forces organized into raid forces.

Attacks and Raids

The GCE may conduct attacks and raids to create situations that will permit seizing and maintaining the political initiative. Successful attacks and raids place considerable pressure on governments and groups supporting terrorism. The decision to execute an attack or raid must include precise identification of the target to ensure that its neutralization will produce the desired political effect. The commander task organizes the force based on METT-T, and the force may include any element, unit, or capability within the MAGTF.

Maritime Interception Operations

Maritime interception operations (MIO) consist of port denial and vessel intercept. Port denial is the act of prohibiting access to specific ports to prevent the import/export of contraband. Vessel interceptions are based on international law associated with maritime visit and search. Boarding parties exercising the right of visit and search may be placed on merchant ships to examine ship's documents, bills of lading, and cargo, searching for evidence of contraband. The GCE may be tasked to provide forces to conduct boarding operations or support to naval special warfare units conducting the intercept.

Peace Operations

There are three distinct types of peace operations—*peacemaking*, *peace enforcement,* and *peacekeeping*.

Types of Peace Operations

Peacemaking. Peacemaking is primarily a diplomatic process beyond the purview of military personnel.

Peace enforcement.   Peace enforcement includes appropriate forceful military actions to separate belligerents involved in a conflict, with or without their consent.   Forces employed in peace enforcement conduct all doctrinal military operations to force a cessation of hostilities.

Peacekeeping operations.   Peacekeeping operations are employed to support diplomatic efforts in order to maintain peace in areas of potential conflict.   The goal, objective, intent, or mission of these operations is to stabilize a conflict between belligerent nations or factions and requires the consent of all parties involved in the dispute.   Peacekeeping operations may be more appropriately referred to as truce-keeping since a negotiated truce is maintained.

The most important requirements for a peacekeeping operation are a negotiated truce and consent to the operation by all the parties in a dispute.   Peacekeeping often involves ambiguous situations requiring the peacekeeping force to deal with extreme tension and violence without becoming a participant.   The essential elements that must be present at the time a peacekeeping force is established, as well as throughout its operation, include the following:

- The consent, cooperation, and support of the authorities of all the parties in the conflict.
- Political recognition of the peacekeeping operation by most, if not the entire, international community.
- A clear, restricted, and realistic mandate or mission with specified and easily understood rules of engagement.
- Sufficient freedom of movement for the force and observers to carry out their responsibilities.
- An effective C^2 system.
- Well-trained and impartial forces.
- An effective and responsive all-source intelligence gathering and dissemination capability.
- Coordinated logistic support of the force.

The GCE may conduct the following missions in support of peace operations—*preventive deployment, internal conflict resolution measures, assistance to (interim) civil authority,* and *protection of humanitarian assistance operations.*

Preventive deployment is the deployment of a multinational force where a conflict threatens to occur. It requires neither a truce nor a peace plan between the potential belligerents. The force deploys at the request of any of the parties involved, without agreement between belligerents except to the specific tasks. The MAGTF's tasks may include—

- Interposing itself in order to forestall violence.
- Protecting the delivery of humanitarian relief.
- Assisting local authorities to protect and offer security to threatened minorities.

Internal conflict resolution measures are the actions taken by a multinational force to restore and maintain an acceptable level of peace and security. They are appropriate to conditions where there is a dispute in which the parties may be less easy to identify than in conventional conflict, and the presence of a multinational force may not enjoy local consent. Although the level of violence may be low and sporadic, the danger to multinational personnel is greater than in a conventional peacekeeping operation because the potential threat is difficult to identify.

Assistance to (interim) civil authority is the task of the multinational force to supervise the provisions of the peace agreement and to ensure that any transfer of power or the holding of elections will be carried out fairly. The role of the military contingents is to maintain a level of security which allows the civil administration to function effectively.

Protection of humanitarian assistance operations entail the employment of a military force to ensure the safe and uncontested delivery of relief supplies and resources. A joint, multinational task force organized for relief protection will need to focus on three main tasks—establishing a mounting base; ensuring delivery of resources; and providing security for the victim population at the delivery site. A multinational relief protection operation may have several of the following characteristics:

- The in-country delivery of relief must be mounted through a secured forward base and not directly from donor nation to victim community.

- Some element of local armed opposition may be encountered, but it is unlikely that it will be of such strength to require intensive combat operations.

- Protection of delivery should be planned with the cooperation of regional neighboring states.

- The response agencies, both nongovernmental organizations and the military force, should be coordinated. This can be done by an organization designed to coordinate the needs of relief agencies with the capabilities of military units.

Operations to Promote Peace

Use of military forces in peacetime helps keep tensions between nations below the threshold of armed conflict and maintains U.S. influence in foreign lands. Such operations are typically joint in nature and may involve forward-deployed MAGTFs, CONUS-based forces, or a combination of both.

Nation Assistance

The U.S. may accelerate security assistance when a friendly or allied nation faces an immediate military threat. Initial support is frequently focused on providing additional combat systems or supplies. MAGTF support to security assistance surges may include the full range of training and logistic support.

Antiterrorism

Antiterrorism is the deterrence of terrorism through active and passive measures. The basics of such a program begin with a well-trained GCE, continuing actions, and security procedures. It also includes the collection and dissemination of timely threat information, the conduct of information awareness programs, personal training, and coordinated security plans. Protective plans and procedures are based on the threat and should strike a reasonable balance between protection, mission requirements, the criticality of assets and facilities, and available manpower and resources. The MAGTF may provide antiterrorism assistance to foreign countries as part of the overall U.S. military foreign internal defense and development programs. This support may include training in bomb detection and disposal; physical security; and the detection, deterrence, and prevention of acts of terrorism.

Support to Counterdrug Operations

Illicit drug trafficking organizations undermine and corrupt regional stability. Because our national security directly depends on regional stability throughout the world, drugs have become a major concern of U.S. foreign policy. The Secretary of Defense's guidance directs an attack on the drug problem in three phases—*at the source, in transit, and in the United States*.

At the Source

The MAGTF GCE may be tasked to assist the counterdrug efforts of cooperating foreign governments, agencies, and forces. This assistance is provided through training and operational support to host-country forces and through technical assistance, intelligence support, and collaboration with host nation law enforcement agencies to prevent the export of illegal drugs.

In-transit

The U.S. military assists in the detection and monitoring of aerial and maritime drug smuggling in-transit to the United States, both outside the CONUS and at the borders and ports of entry of the nation. The GCE can provide ground patrol and surveillance of the border areas and C^2 assets in support of these activities.

In the United States

Marines may provide support to Federal, state, and local law enforcement agencies to include training in planning techniques and procedures, loan of military equipment, logistic support, use of aviation assets, assistance to community antidrug programs, and the use of facilities.

Other Civil Support Operations

These operations encompass worldwide humanitarian assistance, military support to civil authorities, and military assistance for civil disturbances. The GCE, as part of a larger military operation, may assist in disaster relief and support to displaced persons, as well as humanitarian and civic assistance. Units can also augment domestic governments of the United States. Such operations can include support to medical facilities, emergency response, and assistance to law enforcement agencies.

Humanitarian Assistance and Disaster Relief

Humanitarian assistance and disaster relief operations use military personnel, equipment, and supplies to support emergency relief to victims of natural or manmade disasters in the United States and overseas. The GCE can provide or support logistic efforts to move supplies to remote areas, to extract or evacuate victims, and to provide emergency communications, medical support, maintenance, maintenance of law and order, and civil engineering support. Historically, forward-deployed naval forces have been quick to respond to an emergency or disaster. The NEF's inherent flexibility and logistic capabilities make it well suited to provide support to these operations.

Chapter 8

MAGTF GCE Operations in a Joint and Combined Environment

Modern warfare requires the coordination and integration of all U.S. military capabilities to achieve objectives in the face of a wide range of threats. In many types of operations, proper task organization and the integration of all capabilities, often in conjunction with forces from other nations and U.S. agencies, is required to generate decisive joint combat power. The goal is to increase the total effectiveness of the joint force, not necessarily to involve all forces or to involve all forces equally. The joint force commander (JFC) is responsible for integrating these capabilities and contributions in time, space, and purpose, i.e., unity of effort. The MAGTF and its ground combat arm, the GCE, provides one of these capabilities.

The manner in which the JFC organizes his forces directly affects the responsiveness and versatility of joint force operations. The first principle in joint force organization is that the JFC organizes his forces to accomplish the mission based on his vision and concept of operations. Unity of effort, centralized planning, and decentralized execution are key considerations. The JFC may elect to centralize selected functions within the joint force, but should strive to avoid reducing the versatility, responsiveness, and initiative of subordinate forces.

The GCE, as part of a MAGTF operating in a joint and/or combined campaign, provides a unique general-purpose ground capability. It may be employed in a variety of ways in both the littoral and inland areas of the JFC's area of operations. The GCE provides the ground combat capabilities of the MAGTF that tie together naval and continental efforts within the theater.

Service Components

All joint forces include Service components. JFCs may exercise operational control (OPCON) through Service components. This relationship is appropriate when stability, continuity, economy, ease of long-range planning, and scope of operations dictate organizational integrity of Service components. Conducting operations through Service components provides uncomplicated command lines and is the preferred method when the GCE is employed in a joint operation. This permits the employment of the GCE as an integral part of the MAGTF, thus retaining the overall capability of the MAGTF. Responsibilities of the Service component commander include—

- Making recommendations to the JFC on the proper employment of the forces of the Service component.
- Accomplishing such operational missions as may be assigned.
- Selecting and nominating specific units of the parent Service component for assignment to subordinate forces. Unless otherwise directed, these units revert to the control of the Service component commander when such subordinate forces are dissolved.

Functional Components

JFCs may establish functional components to provide centralized direction and control of certain functions and types of operations when it is necessary and feasible to fix responsibility for certain normal, continuing functions or when it is appropriate and desirable to establish the authority and responsibility of a subordinate commander. JFCs may conduct operations through functional components or employ them primarily to manage and coordinate selected functions. The nature of operations, mix of Service forces, and C^2 capabilities are normally primary factors in selecting the functional component commander. Examples of functional component commanders are the—

- Joint force land component commander (JFLCC).
- Joint force maritime component commander (JFMCC).
- Joint special operations component commander (JFSOCC).
- Joint force air component commander (JFACC).

While functional component commanders typically exercise OP-CON over assigned and attached forces and tactical control (TACON) over other military capability or forces made available, JFCs assign missions and establish command relationships to meet the requirements of specific situations. See figure 8-1.

JFCs may also establish a supporting/supported relationship between components to facilitate operations. The MAGTF may be assigned to any functional component as a separate subordinate organization. The JFC must appreciate the capability resident in the MAGTF as a whole and resist the piecemeal assignment of its subordinate elements.

Figure 8-1. National Organization of a Joint Force.

Command Relationships

Unity of effort in joint forces is enhanced through the application of the flexible range of command relationships. Joint force command relationships are an array of options JFCs can use to adapt the organization of assigned forces to situational requirements and arrange component operations in time, space, and purpose. See figure 8-2.

Figure 8-2. Command Relationships.

Combatant Command (Command Authority)

Combatant commanders exercise combatant command (command authority) (COCOM) over assigned forces. This broad authority allows the combatant commanders to perform a variety of functions, including—

- Organizing and employing commands and forces.
- Assigning tasks and designating objectives.
- Giving authoritative direction over all aspects of military operations, joint training, and logistics necessary to accomplish the missions assigned to the command.
- Exercising or delegating OPCON.
- Exercising or delegating authority for logistics.
- Coordinating boundaries with other combatant commanders, U.S. government agencies, or agencies of other countries.

COCOM is exercised only by the CINCs. The commander of the U.S. element of a multinational command can also exercise COCOM only when authorized by the Secretary of Defense. COCOM cannot be delegated.

Operational Control

Operational control (OPCON) may be exercised by commanders at any echelon at or below the level of COCOM (CINC-level). OPCON is inherent in COCOM and is the authority to perform those functions of command over subordinate forces involving organization and employing commands and forces, assigning tasks, designating objectives, and giving authoritative direction necessary to accomplish the mission. OPCON should be exercised through the commanders of subordinate organizations; normally this authority is exercised through subordinate joint forces and Service and/or functional component commanders. OPCON does not, in and of it-self, include authoritative direction for logistics or matters of administration, discipline, internal organization, or unit training. OPCON is the authority to—

- Exercise or delegate OPCON and TACON, establish sup-port relationships among subordinates, and designate coor-dinating authorities.

- Give direction to subordinate commands and forces neces-sary to carry out missions assigned to the command, in-cluding authoritative direction over all aspects of military operations and joint training.

- Prescribe the chain of command within the command.

- Organize commands and forces within the command.

- Employ forces within the command.

- Assign command functions to subordinate commanders.

- Plan, deploy, direct, control, and coordinate the action of subordinate forces.

- Establish plans, policies, and overall requirements for the intelligence activities of the command.

- Conduct joint training and joint training exercises to achieve effective employment of the forces of the command as a whole in accordance with joint doctrine established by the Chairman of the Joint Chiefs of Staff (CJCS).

- Establish an adequate system of control for local defense and delineate such areas of operation for subordinate commanders.

- Delineate functional responsibilities and geographical areas of operation of subordinate commanders.

Tactical Control

Tactical control (TACON) may be exercised by commanders at any echelon at or below the level of combatant command. TACON is the detailed and usually local direction and control of movements or maneuvers necessary to accomplish assigned missions or tasks. TACON provides sufficient authority for controlling and directing the application of force or tactical use of combat support assets. TACON does not provide organizational authority or authoritative direction for administrative and logistic support. The commander of the parent unit continues to exercise those responsibilities unless otherwise specified in the establishing directive. TACON provides the authority to—

- Give direction for military operations.
- Control designated forces, aircraft sorties, or missile launches.

GCE and U.S. Army Integration

Marine Corps and Army forces are structured to operate most efficiently using their own Service doctrine, tactics, techniques, and procedures. Therefore, Marine Corps and Army forces assigned to support the other Service are most effective when unit integrity is maintained.

Interoperability

The interoperability of the GCE and Army ground units is enhanced through effective liaison. As C^2 systems may not be completely compatible, liaison teams can facilitate horizontal coordination between different forces. Differences in combat and combat support capability should be viewed as complementary capabilities. Support rendered by another Service to the GCE should be coordinated by the MAGTF CE to ensure a complete understanding of the support required, the GCE mission, and the procedures to be followed in the conduct of that support.

Transfer Considerations

The GCE and MAGTF commanders should ensure that the authority making the transfer of GCE or Army forces considers the following:

- All subordinate and supporting commanders receive the implementing orders. The implementing order(s) should specify the gaining and losing commands, the unit to be moved, logistic and support arrangements, date-time group the transfer is to be effective, and movement routes.

- The commander of the losing command provides the transferred unit the communications-electronics operating instructions (CEOI), fire support nets, logistic nets, and the linkup recognition signals of the gaining command.

- The commander of the losing command provides the transferred unit sufficient CSS to support itself until new logistic arrangements are established.

- The commander of the gaining command provides unit guides to the tranferred unit. The guides should understand the mission and concept of operations for the transferred unit, locations of logistic support facilities, and the situation in the transferred unit's new sector or zone. The guide team must be able to access the gaining command's fire support, command, and logistic nets. The gaining command must be able to fuel and arm the transferred unit as required. Frequencies, call signs, secure fills, and the gaining unit's CEOIs and standard operating procedures (SOPs) must be provided.

- The tranferred unit moves when and as directed. A liaison team is dispatched to the gaining command as soon as possible after receipt of the transfer order. Communication with the gaining command is established by secure voice prior to linkup.

- Gaining and losing command staffs must coordinate transfer routes, times, linkup points and procedures, and movement control. Pertinent information on the transferred force such as capabilities, limitations, unique systems or requirements, and projected operations must be exchanged. Communications plans and acquisition of necessary support must be arranged.

C^2 Considerations

There are numerous C^2 differences between GCE and Army units in equipment, organization, and procedures. In most cases, staff liaison between collocated, adjacent, and supported/supporting GCE and Army units is required to mitigate these differences. Considerations that must be addressed through liaison include—

- **Frequency management.** Number of frequencies available may be insufficient for the number of required nets.

- **Communications-electronics operating instructions (CEOIs).** System operating instructions within the Army and GCE are structured differently.

- **Communications security (COMSEC).** COMSEC software may be incompatible between the GCE and Army units.

- **Intelligence dissemination.** Collection efforts should be coordinated and information passed between GCE and Army organizations.

GCE-Army Support

Support is the action of a force that aids, protects, complements, or sustains another force. Support relationships may be established between the GCE and Army units to enhance unity of effort, clarify priorities, provide one or the other an additional capability, or combine the effects of similar assets.

General Support

General support is the action given to the supported force as a whole rather than to one of its subordinate units. For example, GCE tank, light armored reconnaissance (LAR), and mechanized units might be placed in general support of the joint force or to an Army force to provide security during initial buildup of land forces in theater. Likewise, an Army multiple launch rocket system (MLRS) unit may be placed in general support of the GCE for a specific phase of an operation.

Direct Support

Direct support is a mission requiring a force to support another specific force and authorizing it to answer directly the supported force's request for assistance. For example, Army field artillery may be placed in direct support of a GCE unit when organic artillery is insufficient to support the commander's scheme of maneuver.

Unless limited by the establishing directive, the commander of the supported force has the authority to exercise general direction of the supporting effort. General direction includes the designation of targets or objectives, timing and duration of the supporting action, and other instructions necessary for coordination and efficiency. The supported commander should consider the accepted tactical practices of the Service of the supporting force. Normally, the supporting commander will prescribe the tactics, methods, communications, and procedures employed by elements of the supporting force.

The establishing authority will dictate the purpose and the scope of the required action. The establishing directive should include—

- The strength of the forces allocated to the supporting mission.
- The time, place, and duration of the supporting effort.
- The priority of the supporting mission relative to the other missions of the supporting force.
- The authority, if any, of the supporting force to depart from its supporting mission in the event of exceptional opportunity or an emergency.

Combined Operations

The strategic goal of collective security and the resulting alliances into which the U.S. has entered require that its armed forces be prepared for combined military operations. Combined operations consist of two of more allied nations employing their forces together to successfuly complete a single mission. The success of a combined operation depends largely on the commander achieving unity of effort within an alliance or a coalition operation. Since no universal doctrine exists for combined warfare, unity of effort can be extremely difficult to accomplish. Individual protocols and contingency plans are developed within each alliance. Coalition operations are even less structured, based on temporary agreements or arrangements. Each combined operation is unique, and key considerations involved in planning and conducting multinational operations vary with the international situation and perspectives, motives, and values of the organization's members. The GCE may operate adjacent to coalition forces, in support of coalition forces, or with coalition forces under GCE control. The GCE commander's awareness of the unique considerations for combined operations enhances his ability to accomplish the GCE mission.

The two types of combined operations—alliance and coalition—are normally determined by the objective and whether the objective or relationship is to be long-term or short-term. An alliance is a result of formal agreements between two or more nations for broad, long-term objectives. Alliance members typically have similar national political and economic systems. NATO is one example. A coalition is an ad hoc arrangement between two or more nations for common action. Coalitions often bring together nations of diverse cultures for a limited period of time. The coalition that defeated Iraq in the 1991 Gulf War is an example. As long as the coalition members perceive their membership and participation as advancing their individual national interests, the coalition can remain intact.

Considerations for Combined Operations

U.S. forces will often be the predominant and most capable force within an alliance or coalition. The GCE can be expected to play a central leadership role, albeit one founded on mutual respect. However, it should not be expected to operate with the MAGTF under other-than-U.S. leadership. Regardless of command relationships, several considerations are germane during the planning and conduct of multinational operations.

National Goals

No two nations share exactly the same reasons for entering an alliance or a coalition. To some degree, participation within an alliance or a coalition requires the subordination of national autonomy by member nations. The glue that binds the combined force is agreement, however tenuous, on common goals and objectives. The GCE commander must strive to understand the different national goals and how these goals can affect conflict termination and the desired end state. Maintaining cohesion and unity of effort requires understanding and adjusting to the perceptions and needs of member nations.

Unity of Effort

Motivations of member nations may differ, but combined objectives should be attainable, clearly defined by the commander or leadership structure of the combined force, and supported by each member nation. Capabilities of each member nation's forces will often differ significantly and must be considered by the JFC when determining the types of missions to be assigned. When combined forces are under the direction of the GCE, the GCE commander should strive to involve all national forces commensurate with their capabilities and to balance this with considerations for national pride, honor, and prestige. The GCE commander should establish a personal, direct relationship with the leaders of other national forces as respect and trust are essential to building and maintaining a strong team.

The GCE commander should include staff members from subordinate combined forces in the decisionmaking process, consistent with the terms established at the founding of the alliance or coalition. Member recommendations should be sought continuously by the GCE commander, but especially during the development of courses of action and ROE, assignment of missions to national forces, and establishment of priorities of effort.

Doctrine, Training, and Equipment

Doctrines, operational competence as a result of training and experience, and types and quality of equipment can vary substantially among the military forces of member nations. At times, national capabilities and national expectations or desires concerning roles to be performed may not be in balance. The commander should seek to optimize the contribution of member forces through training assistance, joint exercises, and sharing of resources.

Cultural Differences

Each member has a unique cultural identity. Even minor differences can have a significant impact on cohesion of the force. The GCE commander should attempt to accommodate—

- Religious holidays and other unique cultural traditions.
- Language differences.
- Dietary restrictions.

Management of Resources

Forces of member nations must be supported either by national assets or through the coalition. Resource contributions will vary between members. Some member nations may contribute logistically while others contribute military forces. Many allied or other friendly militaries are designed for national defense. Operating in their native country, they are self-sustaining. When deployed away from their homeland, however, their combat service support capability diminishes. Consequently, the commander must anticipate this and make provisions for increased support requirements for attached allied units. Furthermore, these requirements may be for nonstandard support, especially in sustenance, ammunition, and medical support.

National Communications

Some member forces will have direct and near immediate communications capability from the operational area to their respective political leadership. This can facilitate coordination but can also be a source of frustration as leadership external to the operational area may issue guidance directly to their deployed forces

.

Other Considerations During Planning and Execution

Additional considerations during planning and execution of multinational operations include—

- Rules of engagement.
- Local law enforcement.
- Command and control.
- Intelligence collection, production, and dissemination.
- Logistics.
- Protection measures such as air defense, defensive counterair, reconnaissance and surveillance, and security measures.

The Campaign

A campaign is a series of related joint major operations that arrange strategic, operational, and tactical actions to accomplish strategic and operational objectives. Within a campaign, major operations consist of coordinated actions in a single phase and usually decide the course of the campaign. Campaigns and major operations can span a wide variety of situations from quick-hitting, limited-objective operations to more extensive or protracted operations.

Marine ground forces will function in a joint or combined campaign in three basic relationships—*a MAGTF operating as an independent command, MAGTF as part of a joint/combined task force,* and *a MAGTF attaching non-Marine elements.*

MAGTF Operating as an Independent Command

First, a MAGTF may operate with other military forces within an area of operations but as an independent command. Coordination will be dictated as necessitated by the adjacency of friendly forces but will not be based on any established command relationship. This situation will most likely occur during the transition from an amphibious operation to subsequent operations ashore.

MAGTF as Part of a Joint/Combined Task Force

The MAGTF, maintaining its organizational integrity, may form part of a larger joint or combined task force. In this case, the MAGTF headquarters will effect most of the coordination and liaison with the non-Marine forces. Again, coordination with non-Marine units will be dictated primarily by the presence of those units adjacent to Marine units.

MAGTF Attaching Non-Marine Elements

The MAGTF may be assigned combat units from other U.S. forces or a foreign military and should, in turn, logically assign those units to the appropriate element when the elements have the capability to absorb them. This relationship most directly involves ground commanders coordinating with non-Marine forces. The commander must consider the requirements and interoperability of communications, fire support, and logistics.

Liaison

The key to successful joint or combined operations is effective liaison at all levels between elements of the different forces. The GCE commander should establish early and frequent liaison to ensure unity of purpose and intent and to standardize procedures. When operating with allied or other friendly forces, the best method of liaison is to exchange liaison teams which are—

- Equipped with mobility assets compatible with the supported unit.
- Equipped with their own organic communications.
- Staffed with sufficient personnel to operate continuously.
- Self-sufficient in equipment and supplies.

Normally, the MAGTF CE will establish the initial liaison with forces of other Services, allied forces, or other friendly forces. It will specify liaison to be conducted by elements of the GCE and the levels at which liaison will take place. Liaison may be required down to the battalion level.

Command Liaison

Commanders of all organizations routinely initiate contact with commanders of other units in their locale even though there may be no official command or support relationship between them. This contact opens the channels of communication which facilitates mutual security, support, and cooperation. This is dictated by common sense as well as by command relationships or by direction from a common superior.

Staff Liaison

Staff officers of all organizations routinely initiate contact with their counterparts at higher, lower, adjacent, supporting, and supported commands. This contact opens channels of communication essential for the proper planning and execution of military operations. Staff liaison may also include temporary assignment of liaison elements to other commands to facilitate continuity of contact and communication. There are three types of liaison elements—*liaison officer, liaison team,* and *couriers.*

Liaison Officer

The liaison officer is the most commonly used technique for maintaining close, continuous contact with another command. A liaison officer is his commander's personal representative and should have the special trust and confidence of the commander to make decisions in the absence of communications. This individual must possess the requisite rank and experience to properly represent his command. Although rank should be a consideration in selection, experience and knowledge of the parent command are more important criteria. A liaison officer often is treated as the duty expert on the employment of his parent unit. The ability to communicate effectively is essential as is sound judgement.

Liaison Team

The liaison team, usually headed by an officer, is assigned when the workload or need for better communications is greater than that within the capabilities of a single liaison officer. The liaison team will normally consist of an officer, liaison chief, clerical personnel/drivers, and communications personnel with equipment. Members of the liaison team may function as couriers as the situation dictates.

Couriers

Although infrequently used due to capabilities of electronic communications, the courier remains a valuable liaison element. The courier is more than a messenger. A courier is dispatched with a specific message and provides information only for a certain period of time. He is expected to provide more information than that contained in the message he is delivering. For this reason, the courier should possess sufficient experience and maturity to respond to questions and provide more than superficial insight into the situation or issues of concern. Individuals selected as couriers will often be staff noncommissioned officers or junior officers. If personnel are available, dedicated couriers may be used to augment the liaison officer or team.

Procedures

Specific techniques and procedures for operations may vary between the GCE and other U.S., allied, or coalition forces. In the conduct of tactical events such as a relief in place, linkup, or passage of lines, the higher commander ordering the operation will specify responsibilities and procedures and will resolve differences in methods of execution. When possible, the commanders of the units involved should be collocated during the operation. When possible, where initial physical contact is established between units, liaison teams should be present. The higher commander must establish measures to ensure continuous and effective fires and other operational support.

The GCE will be required to coordinate directly with forces other than those in the MAGTF whose action in the pursuit of their mission may have an impact on the GCE. Common examples include GCE and non-Marine units operating adjacent to one another; operations requiring a relief in place, passage of lines, or linkup involving the GCE and non-Marine units; or the GCE providing combat support or combat service support to a non-Marine unit, or vice versa. The GCE and other units operating in proximity to one another will be required to coordinate routinely. The most common examples include coordination of cross-boundary fires or movement, exchange of combat information or intelligence, coordination of defensive positions at boundaries, or the coordination of tactical areas of responsibility in counterinsurgency operations. Liaison must take place directly between the elements in contact.

INDEX

Position defense 6-13
Reconnaissance in force 5-9
Strongpoint 6-16

L

Liaison
  Command liaison 8-18
  Couriers 8-20
  Liaison officer 8-19
  Liaison team 8-19
  Joint 8-18
  Qualification 8-19
  Staff liaison 8-19
Light-armored reconnaissance
  General support 8-11
  Reconnaissance 2-3
  Security 2-11
Local security 6-11

M

MAGTF
  Combined arms 1-4
  Commander's intent 1-4
  Joint operations 8-17
Main battle area 6-9
Main battle area forces 6-17
  Locations 6-12
Main echelon
  Current operations section 3-3
  Future operations section 3-4
  Information management 3-6
  Purpose 3-3

Main effort
  Amphibious operations 4-4
  Defense 6-3
  Definition 1-9
  Designation of 2-5
  Enemy 5-7
  Envelopment 5-18
  Exploiting success 5-12
Main effort (Continued)
  Flanking attack 5-17
  Focus of effort 1-11
  Frontal attack 5-15
  Function 1-9
  Main battle area 6-12
  Masking 4-3, 5-8, 5-20, 6-18
  Offense 5-20
  Risk 1-9
  Shifting 1-16, 1-17
  Sustainment 2-7
  Turning movement 5-19
  Weighting 5-1
Maneuver
  Battle damage assessment 2-7
  Close operations 3-14
  Commander's preparation of the
    battlespace 3-6
  Contributors to 1-10
  Defense 6-2
  Definition 1-10
  Description 2-4
  Exploitation of 2-4
  Fire superiority 5-2
  Flexibility 2-5
  Mechanized forces 2-9
  Principle of war 1-10
  Requirements of 2-4
  Violence of action 1-10
Maneuver warfare
  Agility 2-4

U

Unity of command
  OMFTS **4-2**
  Principle of war **1-10**
  Unity of effort **1-10**
Unity of effort
  Combined operations **8-13**
  Command and control
  ~~support 3-5~~
  GCE-Army support **8-11**
  Joint forces **8-5**
  Joint operations **8-1**
  Operations other than war **7-4**
    Impediments to **7-4**

V

Violence **1-17, 2-4, 5-23**
Vulnerability
  Description **1-7**
  Identification of **1-7**
  Relationship to tempo **1-7**

W

Withdrawal
  Amphibious **4-6**
  Purpose **6-21**

Z

Zone of action **1-14**
  Area of operations **3-8**
  Main effort **5-20**
Zone reconnaissance **2-3**

Glossary

I. Acronyms

ACE .... aviation command element
AO .... area of operations
AFFOR .... Air Force forces
ARFOR .... Army forces
atk .... attack

BDA .... battle damage assessment
BHL .... battle handover line

C^2 .... command and control
C^2W .... command and control warfare
CE .... command element
CEOI .... communications-electronic operating instructions
CINC .... commander in chief
CJCS .... Chairman of the Joint Chiefs of Staff
CMO .... civil-military operations
COCOM .... combatant command (command authority)
COMSEC .... communications security
CSS .... combat service support
CSSE .... combat service support element

DLIC .... detachment left in contact

EPW .... enemy prisoner of war

FEBA .... forward edge of the battle area
FLOT .... forward line of own troops
FMF .... Fleet Marine Force

GCE .... ground combat element

IPB .... intelligence preparation of the battlespace

JFACC .... joint force air component commander
JFC .... joint force commander
JFLCC .... joint force land component commander
JFMCC .... joint force maritime component commander
JSOCC .... joint special operations component commander

LAR .... light armored reconnaissance
LC .... line of contact
LD .... line of departure
LOA .... limit of advance

MAGTF .... Marine air-ground task force
MARFOR .... Marine Corps forces
MARFORLANT .... Marine Forces Atlantic
MARFORPAC .... Marine Forces Pacific
MEF .... Marine expeditionary force
METT-T .... mission, enemy, terrain and weather,
                            troops and support available-time available
MIO .... maritime intercept operations
MLRS .... multiple rocket launch system
MOUT .... military operations on urbanized terrain

NATO .... North Atlantic Treaty Organization
NAVFOR .... Navy forces
NCA .... National Command Authorities
NEF .... naval expeditionary force
NEO .... noncombatant evacuation operations

obj .... objective
OMFTS .... operational maneuver from the sea
OOTW .... operations other than war
OPCON .... operational control
PL .... phase line

PPBS .... planning, programming, and budgeting system

R&S .... reconnaissance and surveillance
res .... reserve
ROE .... rules of engagement
RP .... release point
RSTA .... reconnaissance, surveillance, and target acquisition
rte .... route

SOF .... special operations forces
SOP .... standing operating procedure
SP .... start point

TACON .... tactical control
TAOR .... tactical area of responsibility

USAID ...... United States Agency for International Development

II. Definitions

A

amphibious assault — The principal type of amphibious operation that involves establishing a force on a hostile or potentially hostile shore. (Joint Pub 1-02)

amphibious demonstration — A type of amphibious operation conducted for the purpose of deceiving the enemy by a show of force with the expectation of deluding the enemy into a course of action unfavorable to him. (Joint Pub 1-02)

amphibious objective area — A geographical area, delineated in the initiating directive, for purposes of command and control within which is located the objective(s) to be secured by the amphibious task force. This area must be of sufficient size to ensure accomplishment of the amphibious task force's mission and must provide sufficient area for conducting necessary sea, air, and land operations. (Joint Pub 1-02)

amphibious operation — An attack launched from the sea by naval and landing forces, embarked in ships or craft involving a landing on a hostile or potentially hostile shore. As an entity, the amphibious operation includes the following phases: planning, embarkation, rehearsal, movement, assault. (Joint Pub 1-02)

amphibious raid — A type of amphibious operation involving swift incursion into or temporary occupation of an objective followed by a planned withdrawal. (Joint Pub 1-02)

amphibious withdrawal — A type of amphibious operation involving the extraction of forces by sea in naval ships or craft from a hostile or potentially hostile shore. (Joint Pub 1-02)

antiterrorism — Defensive measures used to reduce the vulnerability of individuals and property to terrorist acts, to include limited response and containment by local military forces. Also called AT. (Joint Pub 1-02)

area of influence — A geographical area wherein a commander is directly capable of influencing operations by maneuver or fire support systems normally under the commander's command or control. (Joint Pub 1-02)

area of interest — That area of concern to the commander, including the area of influence, areas adjacent thereto, and extending into enemy territory to the objectives of current or planned operations. This area also includes areas occupied by enemy forces who could jeopardize the accomplishment of the mission. (Joint Pub 1-02)

area of operations — An operational area defined by the joint force commander for land and naval forces. Areas of operation do not typically encompass the entire operational area of the joint force commander, but should be large enough for component commanders to accomplish their missions and protect their forces. (Joint Pub 1-02)

attack — An offensive action characterized by movement supported by fire with the objective of defeating or destroying the enemy. (FMFRP 0-14)

aviation combat element (ACE) — The MAGTF element that is task organized to provide all or a portion of the functions of Marine Corps aviation in varying degrees based on the tactical situation and the MAGTF mission and size. These functions are air reconnaissance, antiair warfare, assault support, offensive air support, electronic warfare, and control of aircraft and missiles. The ACE is organized around an aviation headquarters and varies in size from a reinforced helicopter squadron to one or more Marine aircraft wing(s). It includes those aviation command (including air control agencies), combat, combat support, and combat service support units required by the situation. Normally, there is only one ACE in a MAGTF. (Joint Pub 1-02)

B

battle — A series of related tactical engagements. (FM 100-5)

battle position — A defensive location oriented on the most likely enemy avenue of approach from which a unit may defend or attack. Such units can be as large as reinforced battalions and as small as platoons. The unit assigned to the battle position is located within the general outline of the battle position, but other forces may operate outside the battle position to provide early detection of enemy forces and all-round security. (FMFRP 0-14)

battlespace — All aspects of air, surface, subsurface, land, space, and electromagnetic spectrum which encompass the area of influence and area of interest. (FMFRP 0-14)

blocking position — A defensive position so sited as to deny the enemy access to a given area or to prevent his advance in a given direction. (Joint Pub 1-02)

C

campaign — A series of related military operations aimed at accomplishing a strategic or operational objective within a given time and space. (Joint Pub 1-02)

centers of gravity — Those characteristics, capabilities, or localities from which a military force derives its freedom of action, physical strength, or will to fight. (Joint Pub 1-02)

close operations — Military actions conducted to project power decisively against enemy forces which pose an immediate or near term threat to the success of current battles and engagements. These military actions are conducted by committed forces and their readily available tactical reserves, using maneuver and combined arms. (FMFRP 0-14)

coalition force — A force composed of military elements of nations that have formed a temporary alliance for some specific purpose. (Joint Pub 1-02)

combat service support element (CSSE) — The MAGTF element that is task organized to provide the full range of combat service support necessary to accomplish the MAGTF mission. CSSE can provide supply, maintenance, transportation, deliberate engineer, health, postal, disbursing, enemy prisoner of war, automated information systems, exchange, utilities, legal, and graves registration services. The CSSE varies in size from a Marine expeditionary unit service support group to a force service support group. Normally, there is only one combat service support element in a MAGTF. (Joint Pub 1-02)

combat zone — That area required by combat forces for the conduct of operations. (Joint Pub 1-02)

combatant command — A unified or specified command with a broad continuing mission under a single commander established and so designated by the President, through the Secretary of Defense and with the advice and assistance of the Chairman of the Joint Chiefs of Staff. Combatant commands typically have geographic or functional responsibilities. (Joint Pub 1-02)

combatting terrorism — Actions, including antiterrorism (defensive measures taken to reduce vulnerability to terrorist acts) and counterterrorism (offensive measures taken to prevent, deter, and respond to terrorism), taken to oppose terrorism throughout the entire threat spectrum. (Joint Pub 1-02)

combined arms — The tactics, techniques, and procedures employed by a force to integrate firepower and mobility to produce a desired effect upon the enemy. (FMFRP 0-14)

combined operation — An operation conducted by forces of two or more allied nations acting together for the accomplishment of a single mission. (Joint Pub 1-02)

command — 1. The authority that a commander in the Armed Forces lawfully exercises over subordinates by virtue of rank or assignment. Command includes the authority and responsibility for effectively using available resources and for planning the employment of, organizing, directing, coordinating, and controlling military forces for the accomplishment of assigned missions. It also includes responsibility for health, welfare, morale, and discipline of assigned personnel. 2. An order given by a commander; that is, the will of the commander expressed for the purpose of bringing about a particular action. 3. A unit or units, an organization, or an area under the command of one individual. (Joint Pub 1-02)

command and control warfare — The integrated use of operations security, military deception, psychological operations, electronic warfare, and physical destruction, mutually supported by intelligence, to deny information to, influence, degrade, or destroy adversary C^2 capabilities, while protecting friendly C^2 capabilities against such actions. Command and control warfare applies across the operational continuum and all levels of conflict. Also called C^2W. (Joint Pub 1-02 - CJCS MOP 30, Command and Control Warfare)

command element — The MAGTF headquarters. The CE is a permanent organization composed of the commander, general or executive and special staff sections, headquarters section, and requisite communications and service support facilities. The CE provides command, control, and coordination essential for effective planning and execution of operations by the other three elements of the MAGTF. There is only one CE in a MAGTF. (Joint Pub 1-02)

command post — A unit's or subunit's headquarters where the commander and the staff perform their activities. In combat, a unit's or subunit's headquarters is often divided into echelons; the echelon in which the unit or subunit commander is located or from which such commander operates is called a command post. (Joint Pub 1-02)

commander's intent — A clear, concise statement that defines success for the force as a whole by establishing, in advance of events, the battle or campaign's desired endstate. (FMFM 3)

communications zone — Rear part of theater of operations (behind but contiguous to the combat zone) which contains the lines of communications, establishments for supply and evacuation, and other agencies required for the immediate support and maintenance of the field forces. (Joint Pub 1-02)

concept of operations — A verbal or graphic statement, in broad outline, of a commander's assumptions or intent in regard to an operation or series of operations. The concept of operations frequently is embodied in campaign plans and operation plans; in the latter case, particularly when the plans cover a series of connected operations to be carried out simultaneously or in succession. The concept is designed to give an overall picture of the operation. It is included primarily for additional clarity of purpose. Also called commander's concept. (Joint Pub 1-02)

contingency — A possible future emergency involving military forces caused by natural disasters, terrorists, subversives, or by required military operations. Due to the uncertainty of the situation, contingencies require plans, rapid response capability and special procedures to ensure the safety and readiness of personnel, installations, and equipment. When a contingency occurs it normally creates a crisis. (Proposed change to Joint Pub 1-02)

control — Authority which may be less than full command exercised by a commander over part of the activities of subordinate or other organizations. (Joint Pub 1-02)

counterattack — Attack by part or all of a defending force against an enemy attacking force, for such specific purposes as regaining ground lost or cutting off or destroying enemy advance units, and with the general objective of denying to the enemy the attainment of his purpose in attacking. In sustained defensive operations, it is undertaken to restore the battle position and is directed at limited objectives. (Joint Pub 1-02)

counterterrrorism — Offensive measures taken to prevent, deter, and respond to terrorism. Also called **CT**. (Joint Pub 1-02)

cover — The action by land, air, or sea forces to protect by offense, defense, or threat of either or both. (Joint Pub 1-02)

critical vulnerability — A friendly or enemy capability that is both susceptible to attack and pivotal to that force's success. (Proposed change to Joint Pub 1-02)

D

deception — Those measures designed to mislead the enemy by manipulation, distortion, or falsification of evidence to induce him to react in a manner prejudicial to his interests. (Joint Pub 1-02)

deep operations — Military actions conducted against enemy capabilities which pose a potential threat to friendly forces. These military actions are designed to isolate, shape, and dominate the battlespace and influence future operations. (FMFRP 0-14)

delaying operation — An operation in which a force under pressure trades space for time by slowing down the enemy's momentum and inflicting maximum damage on the enemy without, in principle, becoming decisively engaged. (Joint Pub 1-02)

deliberate attack — A type of offensive action characterized by preplanned coordinated employment of firepower and maneuver to close with and destroy or capture the enemy. (Joint Pub 1-02)

destruction fire — Fire delivered for the sole purpose of destroying material objects. (Joint Pub 1-02)

doctrine — Fundamental principles by which the military forces or elements thereof guide their actions in support of national objectives. It is authoritative but requires judgement in application. (Joint Pub 1-02)

E

engagement — Small, tactical conflicts, usually between opposing maneuver forces. (FM 100-5)

envelopment — An offensive maneuver in which the main attacking force passes around or over the enemy's principal defensive positions to attack those positions from the rear or secure objectives to the enemy's rear. (Proposed change to Joint Pub 1-02)

exploitation — An offensive operation that usually follows a successful attack and is designed to disorganize the enemy in depth. (Joint Pub 1-02)

F

feint — A limited-objective attack involving contact with the enemy, varying in size from a raid to a supporting attack. Feints are used to cause the enemy to react in three predictable ways: to employ his reserves improperly, to shift his supporting fires, or to reveal his defensive fires. (FMFRP 0-14)

fire support coordination — The planning and executing of fire so that targets are adequately covered by a suitable weapon or group of weapons. (Joint Pub 1-02)

firepower — 1. The amount of fire which may be delivered by a position, unit, or weapon system. 2. Ability to deliver fire. (Joint Pub 1-02)

flanking attack — An offensive maneuver directed at the flank of an enemy. (Joint Pub 1-02)

fratricide — The employment of friendly weapons and munitions with the intent to kill the enemy or destroy his equipment or facilities, which results in the unforeseen and unintentional death or injury to friendly personnel. (FM 100-5)

frontal attack — An offensive maneuver in which the main action is directed against the front of the enemy forces. (Joint Pub 1-02)

functional component command — A command normally, but not necessarily, composed of forces of two or more Military Departments which may be established across the range of military operations to perform particular operational missions that may be of short duration or may extend over a period of time. (Joint Pub 1-02)

G

ground combat element — The MAGTF element that is task organized to conduct ground operations. The GCE is constructed around an infantry unit and varies in size from a reinforced infantry battalion to one or more reinforced Marine division(s). The GCE also includes appropriate combat support and combat service support units. Normally, there is only one GCE in a MAGTF. (Joint Pub 1-02)

guard — A security element whose primary task is to protect the main force by fighting to gain time, while also observing and reporting information. (Joint Pub 1-02)

H

harassing fire — Fire designed to disturb the rest of the enemy troops, to curtail movement, and, by threat of losses, to lower morale. (Joint Pub 1-02)

hasty attack — In land operations, an attack in which preparation time is traded for speed in order to exploit an opportunity. (Joint Pub 1-02)

humanitarian and civic assistance — Assistance to the local populace provided by predominantly US forces in conjunction with military operations and exercises. This assistance is specifically authorized by title 10, United States Code, section 401, and funded under separate authorities. Assistance provided under these provisions is limited to (1) medical, dental, and veterinary care provided in rural areas of a country; (2) construction of rudimentary surface transportation systems; (3) well drilling and construction of basic sanitation facilities; and (4) rudimentary construction and repair of public facilities. Assistance must fulfill unit training requirements that incidentally create humanitarian benefit to the local populace. (Joint Pub 1-02)

I

intelligence — The product resulting from the collection, processing, integration, analysis, evaluation, and interpretation of available information concerning foreign countries or areas. (Joint Pub 1-02)

intelligence preparation of the battlespace — An analytical methodology employed to reduce uncertainties concerning the enemy, environment, and terrain for all types of operations. Intelligence preparation of the battlespace builds an extensive data base for each potential area in which a unit may be required to operate. The data base is then analyzed in detail to determine the impact of the enemy, environment, and terrain on operations and presents it in graphic form. Intelligence preparation of the battlespace is a continuing process. Also called **IPB**. (Joint Pub 1-02)

interdiction — An action to divert, disrupt, delay, or destroy the enemy's surface military potential before it can be used effectively against friendly forces. (Joint Pub 1-02)

J

joint operations area — An area of land, sea, and airspace, defined by a geographic combatant commander or subordinate unified commander, in which a joint force commander (normally a joint task force commander) conducts military operations to accomplish a specific mission. Joint operations areas are particularly useful when operations are limited in scope and geographic area or when operations are to be conducted on the boundaries between theaters. Also called JOA. (Joint Pub 1-02)

joint task force — A joint force that is constituted and so designated by the Secretary of Defense, a combatant commander, a subunified commander, or an existing joint task force commander. Also called **JTF**. (Joint Pub 1-02)

L

liaison — That contact or intercommunication maintained between elements of military forces to ensure mutual understanding and unity of purpose and action. (Joint Pub 1-02)

M

main battle area — That portion of the battlefield in which the decisive battle is fought to defeat the enemy. For any particular command, the main battle area extends rearward from the forward edge of the battle area to the rear boundary of the command's subordinate units. (Joint Pub 1-02)

main effort — The designated unit to which is provided the necessary combat power and support, that is designed to successfully attack an enemy vulnerability, critical vulnerability, or center of gravity. (Proposed change to Joint Pub 1-02)

maneuver — Employment of forces on the battlefield through movement in combination with fire, or fire potential, to achieve a position of advantage in respect to the enemy in order to accomplish the mission. (Joint Pub 1-02)

Marine air-ground task force — A task organization of Marine forces (division, aircraft wing, and service support groups) under a single command and structured to accomplish a specific mission. The Marine air-ground task force (MAGTF) components will normally include command, aviation combat, ground combat, and combat service support elements (including Navy Support Elements). (Joint Pub 1-02)

maritime prepositioning ships — Civilian-crewed, Military Sealift Command-chartered ships which are organized into three squadrons and are usually forward-deployed. These ships are loaded with prepositioned equipment and 30 days of supplies to support three Marine expeditionary brigades. (Joint Pub 1-02)

military operations other than war — The range of military actions required by the National Command Authorities, except those associated with major combat operations conducted pursuant to a declaration of war or authorized by the War Powers Limitation Act, in support of national security interests and objectives. These military actions can be applied to complement any combination of the other instruments of national power and occur before and after war. (Proposed change to Joint Pub 1-02)

mission type order — 1. Order issued to a lower unit that includes the accomplishment of the total mission assigned to the higher headquarters. 2. Order to a unit to perform a mission without specifying how it is to be accomplished. (Joint Pub 1-02)

mobile defense — Defense of an area or position in which maneuver is used with organization of fire and utilization of terrain to seize the initiative from the enemy. (Joint Pub 1-02)

mobility — A quality or capability of military forces which permits them to move from place to place while retaining the ability to fulfill their primary mission. (Joint Pub 1-02)

movement to contact — An offensive operation designed to gain or reestablish contact with the enemy. (Joint Pub 1-02 - under "Advance to Contact")

mutual support — That support which units render each other against an enemy, because of their assigned tasks, their position relative to each other and to the enemy, and their inherent capabilities. (Joint Pub 1-02)

N

neutralization fire — Fire which is delivered to render the target ineffective or unusable. (Joint Pub 1-02)

noncombatant evacuation operations — Operations that relocate threatened civilian noncombatants from locations in a foreign country or host nation. These operations normally involve US citizens whose lives are in danger. They may also include selected host nation natives and third country nationals. (FM 100-5)

O

objective — The physical object of the action taken, e.g., a definite tactical feature, the seizure and/or holding of which is essential to the commander's plan. (Joint Pub 1-02)

operational maneuver from the sea — A concept for projecting naval power ashore in support of a strategic objective. Essentially the application of maneuver warfare principles to the maritime portion of a theater campaign, OMFTS capitalizes on the ever expanding capabilities of modern naval and landing forces to project power ashore in an increasingly sophisticated and lethal environment. (OMFTS Concept Paper)

P

position defense — The type of defense in which the bulk of the defending force is disposed in selected tactical localities where the decisive battle is to be fought. Principal reliance is placed on the ability of the forces in the defended localities to maintain their positions and to control the terrain between them. The reserve is used to add depth, to block, or restore the battle position by counterattack. (Joint Pub 1-02 - Also referred to as the "area" defense by the US Army)

power projection — The application of measured, precise offensive military force at a chosen time and place, using maneuver and combined arms against enemy forces. (FMFRP 0-14)

pursuit — An offensive operation designed to catch or cut off a hostile force attempting to escape, with the aim of destroying it. (Joint Pub 1-02)

R

rear area — For any particular command, the area extending forward from its rear boundary to the rear of the area of responsibility of the next lower level of command. This area is provided primarily for the performance of combat service support functions. (Joint Pub 1-02)

rear operations — 1. Military actions conducted to support and permit force sustainment and to provide security for such actions. (FMFRP 0-14) 2. Operations that assist in providing freedom of action and continuity of operations, logistics, and battle command. Their primary purposes are to sustain the current close and deep operations and to posture the force for further operations. (FM 100-5)

reconnaissance in force — An attack designed to discover and/or test the enemy's strength or to obtain other information. (Proposed change to Joint Pub 1-02)

reserve — Portion of a body of troops which is kept to the rear, or withheld from action at the beginning of an engagement, available for a decisive movement. (Joint Pub 1-02)

retirement — An operation in which a force out of contact moves away from the enemy. (Joint Pub 1-02)

retrograde movement — Any movement of a command to the rear, or away from the enemy. It may be forced by the enemy or may be made voluntarily. Such movements may be classified as withdrawal, retirement, or delaying action. (Joint Pub 1-02)

rules of engagement — Directives issued by competent military authority which delineate the circumstances and limitations under which United States forces will initiate and/or continue combat engagement with other forces encountered. Also called **ROE**. (Joint Pub 1-02)

S

scheme of maneuver — The tactical plan to be executed by a force in order to seize assigned objectives. (Joint Pub 1-02)

screen — A security element whose primary task is to observe, identify and report information, and which only fights in self-protection. (Joint Pub 1-02)

sector — An area designated by boundaries within which a unit operates, and for which it is responsible. (Joint Pub 1-02)

security assistance — Group of programs authorized by the Foreign Assistance Act of 1961, as amended, and the Arms Export Control Act of 1976, as amended, or other related statutes by which the United States provides defense articles, military training, and other defense-related services, by grant, loan, credit, or cash sales in furtherance of national policies and objectives. (Joint Pub 1-02)

service component command — A command consisting of the Service component commander and all those Service forces, such as individuals, units, detachments, organizations, and installations under the command, including the support forces that have been assigned to a combatant command, or further assigned to a subordinate unified command or joint task force. (Joint Pub 1-02)

specified command — A command that has a broad, continuing mission, normally functional, and is established and so designated by the President through the Secretary of Defense with the advice and assistance of the Chairman of the Joint Chiefs of Staff. It normally is composed of forces from only one Service. Also called specified combatant command. (Joint Pub 1-02)

spoiling attack — A tactical maneuver employed to seriously impair a hostile attack while the enemy is in the process of forming up or assembling for an attack. (Joint Pub 1-02)

strategy — The art and science of developing and using political, economic, psychological, and military forces as necessary during peace and war, to afford the maximum support to policies, in order to increase the probabilities and favorable consequences of victory and to lessen the chances of defeat. (Joint Pub 1-02)

strongpoint — A key point in a defensive position, usually strongly fortified and heavily armed with automatic weapons, around which other positions are grouped for its protection. (Joint Pub 1-02)

supporting effort — An offensive operation carried out in conjunction with a main effort and designed to achieve one or more of the following: deceive the enemy; destroy or pin down enemy forces which could interfere with the main effort; control ground whose occupation by the enemy will hinder the main effort or force the enemy to commit reserves prematurely or in an indecisive area. (Propsed change to FMFRP 0-14)

suppression — Temporary or transient degradation by an opposing force of the performance of a weapons system below the level needed to fulfill its mission objectives. (Joint Pub 1-02)

T

tactical area of responsibility — A defined area of land for which responsibility is specifically assigned to the commander of the area as a measure for control of assigned forces and coordination of support. Also called **TAOR**. (Joint Pub 1-02)

tactics — 1. The employment of units in combat. 2. The ordered arrangement and maneuver of units in relation to each other and/or to the enemy in order to use their full potentialities. (Joint Pub 1-02)

task organization — A temporary grouping of forces designed to accomplish a particular mission. Task organization involves the distribution of available assets to subordinate control headquarters by attachment or by placing assets in direct support or under the operational control of the subordinate. (FMFRP 0-14)

tempo — The rate of military action; controlling or altering that rate is a necessary means to initiative; all military operations alternate between action and pauses as opposing forces battle one another and fight friction to mount and execute operations at the time and place of their choosing. (FM 100-5)

theater — The geographical area outside the continental United States for which a commander of a combatant command has been assigned responsibility. (Joint Pub 1-02)

turning movement — A form of offensive maneuver in which the attacking force passes around or over the enemy's principal defensive positions to secure objectives deep in the enemy's rear to force the enemy to abandon his position or divert major forces to meet the threat. (Proposed change to Joint Pub 1-02)

U

unified command — A command with a broad continuing mission under a single commander and composed of significant assigned components of two or more Military Departments, and which is established and so designated by the President, through the Secretary of Defense with the advice and assistance of the Chairman of the Joint Chiefs of Staff. Also called unified combatant command. (Joint Pub 1-02)

V

vulnerability
susceptible to attack. (Proposed change to FMFRP 0-14)

W

withdrawal operation — A planned operation in which a force in contact disengages from an enemy force. (Joint Pub 1-02)

www.ingramcontent.com/pod-product-compliance
Lightning Source LLC
Chambersburg PA
CBHW031257090426
42742CB00007B/493